Civil Society Organizations in Latin American Education

Examining the roles, impacts, and challenges of civil society organizations (CSOs) in Latin America, this volume provides a broad perspective on the range of strategies these organizations employ and the obstacles they face in advocating for and delivering educational reform. Building on previous research on international and comparative education, development studies, social movements, and non-governmental organizations, chapter authors provide new insights about the increasing presence of CSOs in education and offer case studies demonstrating how these organizations' missions have evolved over time in Latin America.

Regina Cortina is Professor of Education at Teachers College, Columbia University, USA.

Constanza Lafuente is Adjunct Assistant Professor at Teachers College, Columbia University, USA.

Routledge Research in International and Comparative Education

This is a series that offers a global platform to engage scholars in continuous academic debate on key challenges and the latest thinking on issues in the fast-growing field of International and Comparative Education.

For more information about the series, please visit www.routledge.com.

Books in the series include:

Global Literacy in Local Learning Contexts
Connecting Home and School
Mary Faith Mount-Cors

Dialogue in Places of Learning
Youth Amplified in South Africa
Adam Cooper

Faculty Development in Developing Countries
Improving Teaching Quality in Higher Education
Edited by Cristine Smith and Katherine E. Hudson

Teaching and Learning Difficult Histories in International Contexts
A Critical Sociocultural Approach
Edited by Terrie Epstein and Carla L. Peck

Globalization and Japanese "Exceptionalism" in Education
Insider's Views into a Changing System
Edited by Ryoko Tsuneyoshi

Canadian Teacher Education
A Curriculum History
Edited by Theodore Michael Christou

The Shifting Global World of Youth and Education
Edited by Mabel Ann Brown

The Making of Indigeneity, Curriculum History, and the Limits of Diversity
Ligia L. López López

Civil Society Organizations in Latin American Education
Case Studies and Perspectives on Advocacy
Edited by Regina Cortina and Constanza Lafuente

Civil Society Organizations in Latin American Education
Case Studies and Perspectives on Advocacy

Edited by Regina Cortina
and Constanza Lafuente

First published 2018
by Routledge
711 Third Avenue, New York, NY 10017

and by Routledge
2 Park Square, Milton Park, Abingdon, Oxon, OX14 4RN

Routledge is an imprint of the Taylor & Francis Group, an informa business

© 2018 Taylor & Francis

The right of Regina Cortina and Constanza Lafuente to be identified as editor of this work has been asserted by them in accordance with sections 77 and 78 of the Copyright, Designs and Patents Act 1988.

All rights reserved. No part of this book may be reprinted or reproduced or utilised in any form or by any electronic, mechanical, or other means, now known or hereafter invented, including photocopying and recording, or in any information storage or retrieval system, without permission in writing from the publishers.

Trademark notice: Product or corporate names may be trademarks or registered trademarks, and are used only for identification and explanation without intent to infringe.

Library of Congress Cataloguing-in-Publication Data
A catalog record for this book has been requested

ISBN: 978-1-138-09741-4 (hbk)
ISBN: 978-1-315-10487-4 (ebk)

Typeset in Sabon
by Apex CoVantage, LLC

Contents

Acknowledgments   vii

Introduction: The Role of Civil Society Organizations in Education   1
REGINA CORTINA AND CONSTANZA LAFUENTE

1 *"Only Quality Education Will Change Mexico"*: The Case of *Mexicanos Primero*   18
REGINA CORTINA AND CONSTANZA LAFUENTE

2 *Mexicanos Primero*: Efforts to Account to Parents and Teachers   42
CONSTANZA LAFUENTE AND REGINA CORTINA

3 The Student Movements to Transform the Chilean Market-Oriented Education System   63
CRISTIÁN BELLEI, CRISTIAN CABALIN, AND VÍCTOR ORELLANA

4 The *Círculos de Aprendizaje* Program in Colombia: The Scaling-Up Process   85
LAURA MARÍA VEGA-CHAPARRO

5 Social Advocacy in Neoliberal Times: Non-governmental Organizations in Ecuador's Refugee Landscape   108
DIANA RODRÍGUEZ-GÓMEZ

6 **The Implications of Education Advocacy for Civil Society Organizations**   129
CONSTANZA LAFUENTE

Contributors   153
Index   157

Acknowledgments

Graduate students in the International and Comparative Education Program at Teachers College, Columbia University, participated as research assistants during the development of this book. Special thanks go to Lucía Caumont-Stipanicic, Amanda Earl, and Victoria Hernández.

As editors of this book, we would like to acknowledge with gratitude the financial support we received from Teachers College, Columbia University, in the research phase of the book, and from the Institute of Latin American Studies, Columbia University, in the preparation of the final manuscript. The final version of the book benefited from the skillful editing of Wendy Schwartz.

Introduction
The Role of Civil Society Organizations in Education

Regina Cortina and Constanza Lafuente

Civil Society Organizations in Latin American Education: Case Studies and Perspectives on Advocacy explores the ways that these organizations influence public education policy and individual and collective views about education. Presenting advocacy as a legitimate and effective strategy to influence policy, the six chapters in this volume, all written by experts in the field, highlight the significance of advocacy activities as civil society becomes more robust, experienced, and active in education. The case studies described demonstrate the diversity of the organizations and groups seeking to improve education: from political advocacy to encouragement of citizens' engagement in campaigns, to support for student movements implementing comprehensive strategies involving social protest or litigation, to cooperation with governments by mainstreaming and scaling-up education innovations to shape policy design and implementation.

The State of Education in Latin America

Recent History

During the 1980s, many Latin American countries' economies nearly collapsed. To stem the economic crisis, the next decade brought the implementation of the Washington Consensus, a set of ten neoliberal market-oriented reforms that favored the retrenchment of welfare states. Since the subsequent sharp reduction of public expenditures for education, countries in Latin America have been challenged to allocate the necessary public resources to increase graduation rates, improve academic achievement, reduce the education opportunity gaps for Indigenous groups and rural populations, and provide quality education for all children in public schools.

The Current Situation

The majority of Latin American countries now provide access to free and compulsory primary schooling and have nearly eliminated the gender gap at that level. The Education for All (EFA) Regional Review (UNESCO 2014), a component of UNESCO's global movement to improve the quality of education in developing nations, shows that access to pre-primary and secondary schooling has continued to expand. Half of Latin American countries have a gross enrollment ratio of at least 80 percent in pre-primary education. In 2010, approximately 53.5 percent of Latin American youth aged 20 to 24 had completed their secondary education, compared with the 44.8 percent of youth who completed this level of education in 2000 (UNESCO 2014).

But recent results from the Program for International Student Assessment (PISA), the global education survey conducted by the Organization for Economic Co-operation and Development (OECD), show that despite increased access to education, Latin American students lag behind in performance in comparison to other OECD countries. Half of the students from the ten Latin American countries that participated in PISA in 2015 did not reach the basic achievement level in science, and a high percentage of students demonstrated low performance in reading (46 percent) and math (63 percent) (Bos et al. 2016). Thus, providing the kind of quality of education that would allow students to take up their roles as full citizens in society remains a challenge.

Teachers are key actors in improving the quality of education, but several limitations undermine their effectiveness. For instance, lack of investment in teachers' professional development and the overall quality of their employment have led to low morale and have had a negative impact on Latin American education systems (UNESCO 2009). Further, for a variety of reasons, school systems do not attract the best candidates for teaching positions. Recruitment processes are not sufficiently selective, salaries are not attractive, and the low prestige of the teaching profession makes it unappealing for high achievers (Bruns and Luque 2015). Performance-based teacher evaluation is being developed in a few countries, including Mexico, Chile, Peru, and Ecuador, with mixed results. In many countries, teachers work in under-resourced public schools with inadequate access to basic services like running water and electricity. Further, numerous strikes and conflicts between teachers' unions and governments aimed at improving teacher salaries have resulted in high teacher absenteeism and a reduced number of school days for students (Fernandez 2015).

Public schools are highly unequal in Latin America. There are sharp disparities in resource allocation, school infrastructure, and education. Decentralization reforms in the 1990s, the result of attempts to make education more cost effective and efficient, intensified inequality by increasing the education outcome gap between the poor and the rich. Socioeconomic and geographic inequality goes hand in hand. Low-income groups, and Indigenous peoples in the poorest quintiles who live in underdeveloped and rural regions, attend schools that offer a lower quality education, and they have the lowest achievement and education indicators according to *Education for All* (UNESCO, 2014). Overall, they have higher grade repetition and dropout rates and fewer years of school attainment than more affluent and non-Indigenous students. For example, the Economic Commission for Latin America and the Caribbean (ECLAC 2014) has reported worse education indicators for Indigenous students than for non-Indigenous students in Mexico, Ecuador, and Peru—three of the countries with the largest Indigenous populations in the region. Indigenous youth from ages 20–29 in those three countries had on average two fewer years of schooling than non-Indigenous youth (Cortina 2017).

Civil Society Organizations Working for Education Change

The Increasing Prominence of CSOs

Structural adjustment economic policies, implemented to ensure that countries can service their external debt, supported by international organizations such as the International Monetary Fund (IMF) and the World Bank, along with market-oriented reforms, resulted in the retrenchment of the welfare state during the 1990s. Combined with high macroeconomic volatility, these policies affected low-income families disproportionally. Cutbacks in the state presence in education and health and social services led to the emergence of a parallel non-state institutional network to provide financial assistance and service delivery in support of schools and low-income students and their families (Pagano et al. 2007). A great diversity of CSOs, such as Catholic charities, non-governmental organizations (NGOs), and membership-based grassroots associations, are examples of the groups that emerged and continue today to support low-income students and schools through provision of infrastructure, school materials, scholarships, tutoring services, afterschool activities, or food distribution.

Moving beyond conventional models centered on service provision and financial assistance, a growing number of CSOs concerned about education are now transcending traditional roles to become active participants in policy debates, addressing the root causes of sustained education inequality. These innovative organizations seek to transform civil society's attitudes toward public education and restore the centrality of education in public debates; their goal is to generate national conversations about how to improve policies and institutions. In all countries, they coexist with the many foundations and other types of philanthropic and membership-based organizations that continue to play a role in service provision and assistance for public education.

The work of these CSOs, however, counters the traditional belief of individuals in many countries that the average citizen is not able to participate in public education or does not have a voice to influence education reform due to the unwillingness of public sector authorities and policy makers to incorporate their views into such policy debates. Thus, to illuminate the developing and productive relationship between individuals with needs yet to be met by their governments and the civil society organizations supporting them, the chapter authors in this volume build on research in many related fields, such as comparative education, organizational studies, and international education development, to provide new insights into civil society participation in Latin America. Their underlying premise is that organizations in civil society have intensified their advocacy focus in education to take part in policy debates, influence policy development and implementation, and transform citizens' attitudes and opinions toward public education to guarantee the right to education for all and to extend educational opportunities to those who are yet to fully benefit from them.

Comparative Education Research on Civil Society Organizations

The trajectory of international development organizations is vital for understanding the rising centrality and influence of civil society organizations in the field of education. International organizations propelled the growth of CSOs by selecting them as key stakeholders in their own reform efforts. The 2000 Dakar Framework for Action, which put forth six measurable education goals for meeting the learning needs of all children and which was adopted in the World Education Forum, highlighted the prominence of civil society in the achievement of the *Education for All* targets (UNESCO 2000). Further, an evaluation of national education coalitions' roles confirmed the importance of civil society in achieving EFA's targets by boosting citizen participation, holding governments accountable, and increasing the effectiveness of EFA aid (Global Partnership for Education 2012).

The rising presence of CSOs described in this volume has sparked widespread academic interest in understanding their roles, practices, and impact on education. Within the field of comparative education, research about these organizations emphasizes their service delivery roles (Rose 2009; Sutton and Arnove 2004) or explores national and transnational advocacy networks that influence change in education policy (Archer 2010; Mundy 2008, 2012; Mundy and Murphy 2001; Stromquist 2008; Verger and Novelli 2012). The case studies described in this volume build upon this second line of research by focusing on the education advocacy of CSOs in Latin American countries.

The Themes of This Volume

Below we define and discuss the themes and topics characterizing the six chapters that follow: civil society and civil society organizations, and advocacy, especially education advocacy.

Civil Society and the Nature and Role of CSOs

We understand CSOs as the organizational component of the infrastructure of civil society (Lewis 2007). Cohen and Arato (1992) define this component as the sphere of social interaction between the market and the state, composed of families, associations, social movements, and all forms of public debate. It includes organizations with diverse characteristics and degrees of formalization and access to political and economic resources. Examples of CSOs—grassroots organizations, cooperatives, NGOs, unions, philanthropic foundations, mutual aid associations, religious organizations, community-based organizations, transnational advocacy networks, and student organizations—suggest their diverse structures, ideologies, and resources. In Latin America, the patterns of social participation and social inequality are also mirrored in the accessibility of resources for the different types of organizations.

Improvement of public education is the goal of all the CSOs—located in the Global South with headquarters in Latin America—profiled in this volume. Embodying the operational definition of CSOs in general, they are private, self-governing, nonprofit, and voluntary (Salamon 1994). All of them are formally constituted at least to some extent. They are private since they are distinct from the public sector, self-governing since they control their own activities, nonprofit since they do not distribute profits to their directors, and voluntary because they depend to some degree on voluntary participation.

In some cases, CSOs work alongside student and social movements in pursuit of a common agenda. Social movements are networks of individuals and groups that are defined as the social processes through which various actors with a shared identity allocate resources to engage in collective action through social protest against a distinctive opponent (Diani and Bison 2004). Organizations and movements might collaborate to realize shared purposes to transform the social order or wide-reaching policies, such as the case of CSOs that provided legal counsel to the student movements focused on Chilean education, as described in Chapter 3.

Civil society organizations are by definition autonomous from state entities. Yet, in Latin America, the distinction between these two spheres is often unclear. States and civil society are never inexorably separate because they "are always mutually constitutive" (Dagnino 2010, 26). State institutional mechanisms guarantee the framework for civil society to operate, and civil societies hold governments accountable and encourage citizen participation (Oxhorn 2006).

Since the 1980s, Latin American CSOs have been persistently assuming and then augmenting diverse roles, including those of service provider, agenda setter, innovator, monitor of public policy, and technical expert. Their rising centrality is related to the transition to democracy from military dictatorships in the southern cone in the 1980s and 1990s, and the democratization processes that followed in other countries in the region and the consequent increased freedom of association for civil societies. Closely connected is the diffusion of communications and technology, and the growing number of educated professionals who support the creation of new organizations in areas such as human rights, rural development, education, and health (Salamon 1994). Most importantly, the growth in outsourcing of education and social delivery services by international donor agencies, as well as increased funding by governments to implement such services, explain CSOs' expanding presence. Second-generation reforms implemented in the region, which aimed to decrease public sector intervention in the economy, also encouraged contracting out to CSOs in order to obtain improved transparency of public finances.

Education Advocacy

Our definition of advocacy is consistent with Jenkins' (2006) broad conception: Advocacy comprises actions that seek to influence change in public decision makers, private organizations, and citizens on behalf of the public interest. Such advocacy tends to represent the interests of individuals that political systems or economic structures exclude. Thus, it claims to "represent the collective interests of the general public and underrepresented groups as opposed to the interests of the well-organized powerful groups especially business, mainstream social institutions and the elite professions" (307). The advocacy strategies of these organizations are seen through their activities, such as campaigning to raise awareness about human rights, fostering citizen participation in education, protesting, seeking to influence national or international policy through lobbying public authorities, monitoring education policies or EFA achievement, and initiating and litigating class action suits for education rights. All of these activities are examples of CSOs engaging in advocacy in the field of education.

We conceptualize three main strategy orientations in the education field, locating them along a continuum that goes from conformity to protest (see model included in Table I.1).

This table identifies the basic categories of education strategies, and related actions, that CSOs employ: provision of education services; political, social, and legal advocacy; and social protest. The classifications are supported by theoretical work on advocacy groups in the United States by Minkoff, Aisenbrey and Agnone (2008), Minkoff (1999), and Jenkins (2006). While each action appears as a separate item on the table, it is likely that CSOs combine actions and engage in several at the same time.

The left-hand column of the continuum presents the strategies of organizations that *conform to education models* and do not seek to change policies or institutions since they sanction existing education policies. This is the case of the organizations that deliver education services, distribute goods, or provide assistance.

The middle column identifies approaches of organizations that *seek to reform education through established channels*. Their type of advocacy can be classified as political, social, or legal. Political advocacy targets public decision makers during the different stages of the policy process—for example, agenda setting, implementation, and/or monitoring—and might also

Table I.1 The Strategies of Education of Civil Society Organizations

Provision of Education Services	Political, Social, and Legal Advocacy	Social Protest
Conforms to Education Models	Seeks to Reform Education Through Established Channels	Seeks Transformation Through Unconventional Channels
Sponsorship of schooling by covering direct costs through scholarships Education provision (with or without government funding) Social assistance Tutoring services Infrastructure provision	**Social Advocacy** Campaigns that raise awareness of the right to education Campaigns that encourage citizens' participation in education Efforts to improve the practices of other CSOs **Political Advocacy** Lobbying legislators Technical assistance for policy design and implementation Policy monitoring Monitoring achievement of EFA targets **Legal Strategy** Strategic litigation Class action suits	Protest Boycotts

include legal tactics that target the judiciary such as strategic litigation and class actions. Social advocacy is instead directed toward private actors such as organizations, citizens, and businesses, and it aims to modify their viewpoints about education policies and practices.

The right-hand column of the continuum refers to advocacy strategies that operate outside of conventional institutional channels and *seek to transform through unconventional channels*. Such strategies aim to replace or radically transform political, economic, and institutional mechanisms that generate education inequities. Such an approach based on social protest is characteristic of social movements.

The Contents of This Volume

Chapter 1: Developing and Refining an Impact Strategy

Here, through a qualitative study, Regina Cortina and Constanza Lafuente explore the political, social, and legal advocacy of *Mexicanos Primero*.

Mexicanos Primero is one of the first civil society organizations in Mexico to demand accountability from federal and state authorities for governmental expenditure of public resources targeted to public education. The authors investigate how *Mexicanos Primero's* strategy combines different approaches to fulfill its organizational mission to increase educational equity and improve education outcomes. The authors develop three arguments based on their investigation. The first contends that *Mexicanos Primero* is part of a broader context in which Mexican civil society has continued to grow and diversify. The second argument asserts that research on corporate involvement in education has great explanatory value for understanding this organization, since *Mexicanos Primero* is an organization that receives a new type of support: funding from corporate sector leaders for education that, contrary to traditional patterns of philanthropy, seeks to actively encourage policy change in education. The third argument contends that a strong organizational capacity, financial resources, and access to decision makers strengthen the effectiveness of *Mexicanos Primero's* strategic advocacy framework.

Such a strategic advocacy framework includes the simultaneous and mutually reinforcing combination of five components: a carefully designed impact policy agenda that details the changes in education policy that will ensure achievement of a fairer education system; applied research to provide solutions to education problems and to support its claims; multimedia campaigns to raise awareness in citizens to generate support for the organization's petitions; professional awards and capacity-building initiatives for teachers and school communities; and a legal approach to put pressure on the public authorities to invest in school infrastructure, and to stop abuses of power from the government and teachers' unions.

Chapter 2: Including Parents, Teachers, and Students

This chapter, also written by Constanza Lafuente and Regina Cortina, explores the accountability challenges of *Mexicanos Primero*, highlighting the tensions in the accountability practices that sustain its advocacy strategies. The authors critically examine how the CSO is accountable primarily to parents and teachers, two of the main groups affected by the problems of education in Mexico and, therefore, targets of its campaigns; and how *Mexicanos Primero* has designed and implemented accountability mechanisms to meet the needs of a broad range of stakeholders.

As advocacy organizations have become notable actors petitioning for education reform and teacher evaluations, their practices supporting advocacy have also become major topics of study, as have their external and internal dimensions of accountability. While the former includes reporting to multiple actors in their ecosystems, the latter dimension refers to CSOs' adherence to their values and mission. In discussing both dimensions of accountability as they apply to *Mexicanos Primero*, the authors provide comparative education researchers with a blueprint for investigating the accountability practices and outcomes of CSOs who are increasingly involved in advocacy work.

The authors argue that the continuous and pressing need to be accountable to donors partly explains why *Mexicanos Primero's* accountability practices prioritize donors over teachers and parents. Three elements from the CSO accountability literature help to explain *Mexicanos Primero's* accountability practices: the need to adhere to its values and the imperative to ensure mission accomplishment; the need to prioritize stakeholders that are key to mission fulfillment; and the presence of remote stakeholders, linked to its status as a non-membership based organization. The authors conclude that there are tensions between the values and principles that the organization espouses—the right of all Mexicans to a quality education—and its accountability practices, which increase the distance between the organization and the groups in civil society that receive an education of a lower quality.

Chapter 3: Collaborating With Student Movements

This chapter provides an in-depth examination of the implementation of market-oriented education reforms in Chile and the opposition to these reforms led by critical and powerful student movements that shook Chilean society by rejecting the rule of the market dynamics. Two student movements situated in the post-Pinochet regime are covered: the "Penguin Revolution" in 2006, led by high school students, and the "Chilean Winter" in 2011, led by university students. This qualitative study demonstrates how these movements became highly relevant political actors in the educational arena. The chapter authors, Cristián Bellei, Cristian Cabalin, and Víctor Orellana, illustrate the movements' education advocacy approaches and identify student organizations as the key actors leading both movements. Many student organizations across the country participated in these movements, which also had the support of several CSOs, including teachers' unions, Chilean development NGOs, and other university student organizations. The study examines the types of supports that CSOs provided to the movements, such as guidance on how to present their ideas during the legislative process and how to produce programmatic documents.

The authors analyze the main features of the Chilean educational system during this time, including its extreme degree of marketization, which provided the institutional context for the formation of the two movements and their key characteristics. They describe the basic features of each, identifying their common elements from an education policy perspective, and focus on the links between students' demands and discourses, and the market-oriented institutions that prevail in Chilean education. The authors also identify the movements' impact on educational debates and policies in Chile, and examine the complexities in implementing the necessary changes to reverse some of the marketization dynamics. The chapter concludes by reminding us that although the relevance of international organizations in the education policy field is ever increasing, policy makers are still socially and locally accountable to civil society at the national level.

Chapter 4: Scaling-Up and Education Reform

In this chapter, Laura María Vega-Chaparro examines the scaling-up of the *Círculos de Aprendizaje* (Learning Circles) program, an educational innovation originally designed by *Fundación Escuela Nueva Volvamos a la Gente* (FEN). *Círculos de Aprendizaje* is a cost-effective educational program that offers quality primary education in rural multigrade schools for internally displaced children who have been out of the educational system for at least six months, and seeks to reintegrate them into the educational system. The author examines the mainstreaming of the program due to its excellent results in the pilot phase by the Colombian Ministry of Education, delving into the challenges of scaling up such an intervention at the national level.

In this qualitative study, the author investigates the *Círculos de Aprendizaje* program as a case of political advocacy focused on education policy. Identifying the most important lessons of this case study, Laura Vega contends that while *Círculos de Aprendizaje* is a good example of how an education innovation designed by a CSO influences education policy, mainstreaming and scaling-up are nevertheless not straightforward processes. She highlights the conditions under which education programs can be implemented at scale, with a focus on sustainability, local ownership, and quality interventions that effectively change practices in schools and classrooms. Finally, she discusses the complexities of the roles of the Ministry of Education and *Escuela Nueva*, and concludes that the participation of the Ministry as a technical advisor, along the scaling-up process, could have substantially improved the quality and sustainability of the initiative.

Chapter 5: Advocacy in Neoliberal Times

Here, Diana Rodríguez-Gómez examines the role of non-governmental organizations (NGOs) as education stakeholders and policy and social advocates in the context of neoliberalism. In this qualitative study, she critically examines NGOs' advocacy, traditionally portrayed as a heroic endeavor, through the lens of the organizations' competitiveness for the funds they need to continue operating in a market of finite economic resources. Specifically, the chapter provides an analysis of the material conditions and social relations of NGOs working in the refugee-aid arena in Ecuador, and illustrates how social advocacy efforts are reworked in order to become more attractive to funding agencies.

To demonstrate how social advocacy projects become marketable, Rodríguez-Gómez uses ethnographic research, including participant observations and interviews, focused on NGO-funded programs committed to providing education services to those in need of international protection. She finds that some NGOs aligned their missions and programs with those of their donors to increase their chances for future funding in the refugee field. Other organizations adopted risk-averse social advocacy tactics by, for example, raising awareness of the situation of young refugees, although without raising any political issues that could be controversial and thus detrimental to their fundraising success.

Chapter 6: The Organizational Practices of Education CSOs

This chapter illuminates the organizational implications of education advocacy for CSOs, determining that robust organizations, guided by appropriate missions and values, are better able to advance improvements in education for all than are less solid organizations. Although comparative education research has increasingly explored education advocacy, it has not examined the implications for organizations of adopting this strategy. Constanza Lafuente draws on the original case study research and findings in the studies of CSOs presented in earlier chapters of this volume to illustrate a rich understanding of an organizational strategy that highlights CSOs' agency. Further, she proposes a framework for conceptualizing the operational environments of these organizations.

This last chapter identifies and describes six practices that support education advocacy: developing cooperative relations with other groups in the organizational environments of CSOs; designing organizational missions in support of education advocacy and aligning programs and accountability with them; combining service delivery with advocacy through scaling-up; acquiring the technical skills and resources required to support advocacy; diversifying resource generation strategies to enhance advocacy sustainability; and obtaining organizational leadership support.

Conclusion

The Importance of CSOs

The growing presence of CSOs advocating for public education demonstrates that, in Latin America, the state is no longer able to support the expansion and the quality of education services that most children receive on its own. These organizations are vital to the expansion of educational services that the state does not provide. The case studies presented in this volume show that Ministries of Education need to build upon the strengths and resources of all social actors, including the multitude of CSOs, and collaborate with both the public and private sectors to realize the right of all to a quality education. It is essential for nation states and civil society organizations to cooperate.

Civil Society Organizations in Latin American Education: Case Studies and Perspectives on Advocacy provides blueprints of the many advocacy approaches through which CSOs are trying to create new spaces of interaction in order to lobby actively in the policy debates on how to improve education quality and equity. Indeed, the case studies demonstrate that advocacy comprises a wide repertoire of actions, some targeted to citizens and private groups, others to public actors or the executive, legislative, and judiciary branches of government. As shown in Chapter 1, CSOs are able to fit their actions into a coherent strategic framework so that they all interact to meet the goal of improving education.

Building relationships, networking, and working together with the public and private sectors and with members of the civil society are crucial elements of political and social advocacy. Further, as examined in Chapter 3, there are additional advocacy strategies, such as protest and mobilization, followed by the articulation and presentation of policy proposals to legislators, which are the main activities of the student movements, and which CSOs have supported (see Table I.1). It is evident, therefore, that in order to advance the right to education, advocacy activists must be strategic about both their activities and their organizational partners.

Recommendations

Collaboration

The case studies demonstrate that initiating and maintaining strong collaboration between organizations and citizens is crucial, so that organizations do not become technical experts detached from the segments of the population that are affected by public education's shortcomings. For example, by assisting the student movements in substantiating their proposals to legislators, CSOs in Chile supported broader social and political presence rooted in citizen participation and strengthened the student movements' framing of education as a social right.

Outreach to Families and Teachers

Issues of public school access and quality are key concerns of students and parents, and teachers are crucial actors in the daily education services provided and the improvement of the learning experience. Therefore, civil society organizations that conduct advocacy need to devise accountability strategies that elicit the participation of different groups to support their initiatives. Particularly, calling upon practitioners for their creativity, reflection, and ability to make adjustments in daily operations will result in organizational learning and ultimately will strengthen the impact of CSOs as they incorporate new perspectives into their work. In some cases, the need to prioritize stakeholders that are key to mission accomplishment may limit the organization's ability to reach out to parents, students, and teachers since organizations necessarily invest more time in reporting to donors than to parents or teachers. Nonetheless, such undertakings are worth the effort.

Mainstreaming and Scaling-Up

As seen in Chapter 4, program mainstreaming and scaling-up are two of the major undertakings that support CSOs' education advocacy because they seek to influence education policy design and implementation. Scaling-up approaches build on the operational and service-related expertise of CSOs by using the evidence of their services to replicate their programs in collaboration with other CSOs, international organizations, Ministries of Education, or regional districts. Mainstreaming education incorporates a CSO's programs into the government's initiatives, representing one of the ways in which replication may occur. This chapter demonstrates that access to policy makers, organizational legitimacy, and data that evidences success of an education innovation in generating qualitative outcomes, all help fulfill a goal of mainstreaming, although such advantages are insufficient to ensure quality outcomes and local ownership of programs once they are implemented at scale. Thus, advocacy approaches require that practitioners in civil society organizations take into account the factors that led to success in the mainstreaming stage and to study their sustainability, local ownership, and quality interventions that effectively change practices in schools and classrooms.

Fundraising

Chapter 5 shows that education advocacy does not happen in an organizational vacuum, but rather it is undertaken by organizations that must compete with each other for scarce resources. Thus, as discussed in Chapter 6, as competition for resources continues to increase in the field of international educational development, CSO practitioners will need to diversify and augment their sources of income to increase both their sustainability and capacity to innovate.

As discussed in Chapters 1 and 2, while cooperating with corporate leaders facilitates access to much-needed resources and capabilities that strengthen education advocacy, it is also true that such collaboration might undermine a CSO's legitimacy as an education advocate. Even though *Mexicanos Primero's* mission statement stresses that public education is a common responsibility of all social actors, in some cases, this organization might need to increase its distance from the citizens and groups that oppose corporate donors to support public education in order to ensure its sustainability. Since traditional labor unions and other sectors see corporate support as a privatization effort, ties to corporate funding can lead to tensions with central groups in civil society whose support organizations need to advance their advocacy goals.

Agenda for Further Research

The case studies presented in this book contribute to building an agenda in comparative education research that addresses education advocacy. First, it is crucial to continue investigating the links between CSOs and social movements. Although CSOs may engage in social protest as movements do, each entity tends to have different strategies. Therefore, research could address, for example, the tensions in such collaboration, how these alliances influence the advocacy approaches of organizations, and what lessons can be learned from these collaborative experiences.

Research is also needed to elucidate how education advocacy addresses the various stages of policy implementation, given that advocacy activities may differ at different stages. For example, research could study CSO involvement in more than one stage, focusing, for example, first on the conditions that affect success in the design and mainstreaming stages, and then a comparison of those conditions with the outcomes for success in the implementation at scale and monitoring stages. Such inquiry could also include a view of the different tactics through which an organization's mainstreaming and scaling-up approaches compensate for the deficiencies in the organizational capacities of Ministries of Education and school districts to implement programs at scale.

Comparative researchers could also identify partnerships between CSOs and those corporate entities that do not support privatization of education. More evidence is required in Latin America to understand the conditions under which the private sector can support advocacy whose prime precept is education as a social right, the nature of the capacities these private actors bring to advocacy organizations, and the shortcomings and tensions inherent in such collaboration.

Research as well can deepen our understanding of the effectiveness of legal strategies in education in Latin America. Comparative studies could identify the conditions under which litigation and appeals are effective in generating concrete advances in the right to education in the region and the nature of the organizational capabilities that support such legal approaches.

Future studies also need to examine how CSOs can mediate tensions when simultaneously engaged in service delivery and education advocacy, given that each undertaking is supported by different funding mechanisms and evaluation methods. Pursuing two different strategies at the same time might severely challenge the management of an organization. Funding available for services tends to be restricted to the delivery of specific education services, whereby organizations need to produce information about performance through quantifiable outcomes. Unrestricted funding instead better supports education advocacy; it allows a CSO to decide freely how to invest its resources to accomplish its advocacy goals. Donors prefer clearly quantifiable interventions that show results within a specific period of time. Funding flexibility, however, is particularly important because advocacy approaches tend to require long-term interventions, and their impact are more difficult to measure. Thus, organizations that are involved in both service delivery and advocacy need to learn to navigate the organizational implications of both strategies.

Future inquiries might examine the strategies, practices, and impact of education civil society organizations by applying the perspective of CSOs *as organizations*. Studies on organizational theory and on nonprofit organizations in particular can shed light on how a CSO's type, size, and structure impact the strategies it implements. Incorporating the organizational perspective can also assist researchers in better understanding the various aspects of advocacy. Research on how organizations operate under conditions of competition for resources can clarify how CSOs' advocacy is underpinned by that competition. Analysis of how CSO accountability mechanisms differ according to organizational sub-types—for example, membership-based CSOs compared with organizations that serve external groups—can help to explain why many accountability efforts fall short of their intentions. Moreover, research on the effect of organization size on capacities can help explain the challenges and relevance of networked structures to increase advocacy engagement, such as national and transnational networks.

Finally, investigating the education advocacy of small CSOs is a topic that deserves further attention since research tends to focus on the advocacy of well-resourced and larger organizations. For their smaller counterparts, advocacy work is harder because sustained advocacy requires highly knowledgeable and trained personnel dedicated full time to that activity, which they are unlikely to have. Studies could identify how smaller CSOs engage in education advocacy by studying their participation in coalition-building and network participation, as well as other practices that support their advocacy engagement.

The editors and chapter authors of *Civil Society Organizations in Latin American Education: Case Studies and Perspectives on Advocacy* anticipate that our book will have an impact on generating new research that examines the growing centrality of civil society in achieving the right to universal education in Latin America. Even though by no means exhaustive on the topic, this volume illuminates new approaches being developed by civil society organizations in Latin America. As these organizations become influential actors in improving public education by expanding educational opportunities for those currently underserved, it is hoped that the accrued benefits will be reflected in the citizenry at large.

References

Archer, David. 2010. "The Evolution of NGO-Government Relations in Education: Action Aid, 1972–2009." *Development in Practice* 20(4/5): 611–618.

Bos, Maria Soledad, Alison Elias, Emiliana Vegas, and Pablo Zoido. 2016. *PISA America Latina y el Caribe*. Washington, DC: Banco Interamericano de Desarrollo.

https://publications.iadb.org/bitstream/handle/11319/7993/Latin-America-and-the-Caribbean-in-PISA-2015-How-Many-Students-are-Low-Performers. PDF?sequence=4

Bruns, Barbara, and Javier Luque. 2015. *Great Teachers: How to Raise Student Learning in Latin America and the Caribbean.* Washington, DC: The World Bank.

Cohen, Jean L., and Andrew Arato. 1992. *Civil Society and Political Theory.* Cambridge, MA: MIT Press.

Cortina, Regina. 2017. *Indigenous Education Policy, Equity, and Intercultural Understanding in Latin America.* New York: Palgrave Macmillan.

Dagnino, Evelina. 2010. "Civil society in Latin America: Participatory Citizens or Service Providers?" In *Power to the People? (Con-) Tested Civil Society in Search of Democracy*, eds. Heidi Moksnes and Mia Melin. Uppsala: Uppsala University: 23–29.

Diani, Mario and Ivano Bison. 2004. "Organizations, Coalitions, and Movements." Special Issue: Current Routes to the Study of Contentious Politics and Social Change. *Theory and Society* 33(3/4): 281–309.

Economic Commission for Latin America and the Caribbean (ECLAC). 2014. *Los Pueblos Indigenas de América Latina: Avances en el Último Decenio y Retos Pendientes para la Garantía de sus Derechos.* Santiago, Region Metropolitana, Chile: ECLAC.

Fernandez, Marco. 2015. "Managing the Politics of Teacher Reform." In *Great Teachers. How to Raise Student Learning in Latin America and the Caribbean*, by B. Bruns and J. Luque. Washington, DC: The World Bank.

Global Partnership for Education. 2012. *Regional and National Civil Society Education Funds—CSEF.* Evaluation Report. Washington, DC: Global Partnership for Education. www.campaignforeducation.org/docs/csef/CSEFev_FINAL_REPORTv4_complete.pdf

Jenkins, Craig. 2006. "Nonprofit Organizations and Political Advocacy." In *The Nonprofit Sector: A Research Handbook*, ed. R. Steinberg and W. Powell. New Haven: Yale University Press.

Lewis, David. 2007. *The Management of Non-Governmental Development.* New York: Routledge.

Minkoff, Debra. 1999. "Bending with the Wind: Strategic Change and Adaptation by Women's and Racial Minority Organizations." *American Journal of Sociology* 104(6): 1666–1073.

Minkoff, Debra, Silke Aisenbrey and Jon Agnone. 2008. "Organizational Diversity in the US Advocacy Sector." *Social Problems:* 55(4): 525–548.

Mundy, Karen. 2008. "From NGOs to CSOs. Social Citizenship, Civil Society and Education For All." *Current Issues in Comparative Education* 10: 32–40.

Mundy, Karen. 2012. "The Global Campaign for Education and the Realization of Education for All." In *Campaigning for "Education for All": Histories, Strategies and Outcomes of Transnational Social Movements in Education*, ed. A. Verger, and M. Novelli. Rotterdam, Netherlands: Sense Publishers.

Mundy, Karen and Lynn Murphy. 2001. "Transnational Advocacy, Global Civil Society? Emerging Evidence from the Field of Education." *Comparative Education Review* 45(1): 85–126.

Oxhorn, Phillip. 2006. "Conceptualizing Civil Society from the Bottom Up: A Political Economy Perspective." In *Civil Society and Democracy in Latin America*, ed. R. Feinberg, C. H. Waisman, and L. Zamosc. New York: Palgrave MacMillan.

Pagano, Ana, Paula Costas, and Ingrid Sverdlick. 2007. *Participación e Incidencia de la Sociedad Civil en las Políticas Educativas. El Caso Argentino.* Foro Latinoamericano de Políticas Educativas. Buenos Aires, Argentina: Laboratorio de Políticas Públicas.

Rose, Pauline. 2009. "NGO Provision of Basic Education: Alternative or Complementary? Service Delivery to Support Access to the Excluded?" *Compare: A Journal of Comparative and International Education* (2): 219–233.

Salamon, Lester. 1994. "The Rise of the Nonprofit Sector." *Foreign Affairs* 73(4): 109–122.

Stromquist, Nelly. 2008. "Revisiting Transformational NGOs in the Context of Contemporary Society." *Current Issues in Comparative Education* 10(1/2): 41–44.

Sutton, Margaret, and Robert Arnove, eds. 2004. *Civil Society or Shadow State? State/NGO Relations in Education.* Greenwich, CT: Information Age.

UNESCO. 2000. *Dakar Framework for Action.* Paris: UNESCO. http://unesdoc.unesco.org/images/0012/001211/121147e.pdf

UNESCO. 2009. *Education for All Global Monitoring Report. Regional Overview: Latin America and the Caribbean.* Paris: UNESCO. http://en.unesco.org/gem-report/sites/gem-report/files/178428e.pdf

UNESCO. 2014. *Latin America and the Caribbean. Education for All 2015 Regional Review.* Paris: UNESCO. http://unesdoc.unesco.org/images/0023/002327/232701e.pdf

Verger, Antoni, and Mario Novelli, eds. 2012. *Campaigning for "Education for All": Histories, Strategies and Outcomes of Transnational Social Movements in Education.* Rotterdam, Netherlands: Sense Publishers.

1 "Only Quality Education Will Change Mexico"
The Case of *Mexicanos Primero*

Regina Cortina and Constanza Lafuente

In this chapter, we analyze the advocacy strategies of *Mexicanos Primero* (Mexicans First), one of the first civil society organizations (CSOs) in Mexico to demand accountability from federal and state authorities for the use of public resources targeted for public education. We investigate how *Mexicanos Primero* combines different activities to fulfill its organizational mission, and conceptualize it as a new type of education CSO: highly professionalized, with a robust organizational capacity and corporate leadership support for education reform in Mexico, an area of public policy that corporate leaders have traditionally left to the state and trade unions.

We present three arguments about the approaches and operations of *Mexicanos Primero* based on our research. First, we assert that this CSO is part of a broader process in which Mexican civil society—a concept that we define as the sphere of social interaction between the market and the state, comprised of voluntary associations and social movements (Cohen and Arato 1992)—has continued to expand and diversify (Verduzco 2003). Second, we argue that although *Mexicanos Primero* is not a corporate foundation and does not claim to represent corporations, utilizing the concepts and scholarship on corporate involvement in education has great explanatory value for this particular case. From this perspective, funding for *Mexicanos Primero* represents a new type of support from corporate leadership for education; contrary to traditional patterns of philanthropy, supporters of *Mexicanos Primero* seek instead to actively encourage policy change in education. With funding provided by individual entrepreneurs, *Mexicanos Primero* employs strategies starkly different from those sanctioned by mainstream corporations involved in education; for example, it does not support vouchers, charter schools, or the privatization of the public school system. Third, we argue that strong organizational capacity, financial resources, and access to decision makers within the Mexican government strengthen the effectiveness of *Mexicanos Primero's* strategic advocacy framework. We use this framework as a mechanism for analyzing the repertoire of activities through which the organization fulfills its organizational purpose, and contend that its effectiveness depends on a simultaneous and mutually reinforcing combination of a series of strategic components.

The chapter is organized as follows: This first section presents the data sources and describes the data-gathering processes. The second describes the origins of *Mexicanos Primero*. The third explains its organizational structure. The fourth section contextualizes our study's focus on advocacy among CSOs. The fifth section analyzes the interrelated components in *Mexicanos Primero's* strategic advocacy framework, and the sixth discusses the three arguments, citing examples of *Mexicanos Primero's* strategies and activities to bolster our conclusions. Finally, the last section reviews and summarizes our findings.

Data Sources

The qualitative research reported in this chapter is based on our analysis of primary and secondary data sources. In June 2015, we interviewed seven management and staff members of *Mexicanos Primero*, including the president, the executive director, five program directors, and one former senior staff member of *Mexicanos Primero*. The president and the executive director facilitated access to management and staff interviewees. Interviews averaged 80 minutes and were all semi-structured and guided by an interview protocol built upon the conceptual framework presented in this chapter. We triangulated data obtained through interviews with secondary data that included annual reports and research reports produced by *Mexicanos Primero's* research team, and relevant newspaper articles. All of the quotes presented in this chapter are taken from interview transcripts.

The Origins of *Mexicanos Primero*

Mexicanos Primero is a CSO created in 2005 as a nonprofit association with the motto "Only quality education will change Mexico."

The Founders and Founding of Mexicanos Primero

Mexicanos Primero was started by a group of prominent business leaders in Mexico City and several university professors who shared an interest in education and public policy. Several of its founding members participated in Mexican philanthropic organizations, such as *Teletón*, a nonprofit organization created in 1997 to raise funds through TV programs for children with special needs and children's rehabilitation centers; and *Fundación Televisa*, the corporate foundation of the mass media company Televisa Group. *Mexicanos Primero's* president was also the president of the Televisa Foundation and the co-founder of *Bécalos*, a scholarship program implemented in collaboration with the Televisa Foundation and the National Association of Mexican Bankers. *Bécalos* provides scholarships and grants to teachers, school leaders, and students by requesting a small donation each time someone uses an ATM. Between 2006 and 2014, 221,323 scholarships were awarded, and many teachers and school principals have benefited from the scholarships (Bécalos 2016). While such philanthropic initiatives represent conventional models of voluntary activity that do not seek policy reform (Minkoff 1999), participating in these initiatives introduced both the president and founders of *Mexicanos Primero* to the problems and challenges of public education in Mexico.

Two of *Mexicanos Primero's* founding members decided to gather colleagues and friends to create an organization that would improve the quality of public education in Mexico. While visiting a school they had helped to build, they shared their concerns about education in Mexico and agreed that more had to be done in the country than helping to build schools or distributing scholarships to low-income students. One of these two founding members was completing a master's degree at the time and contacted one of his professors, who also became interested in the initiative and later joined the new organization as executive director.

The Founding Principles of Mexicanos Primero

During the initial development period, the founding group outlined the main approach for *Mexicanos Primero*. The group believed that the time had come in Mexico to demand structural changes in education and that this new organization could propose such reforms informed by applied research in order to make policy proposals to authorities. The executive director explained the choice they faced between two different orientations:

> There were two possible strategies for the new organization: doing philanthropic work and assistance-related activities, or, instead, creating policy proposals for education. We agreed that we did not want to distribute computers or scholarships to low-income students and schools, but, rather, we wanted to conduct applied research to solve the problems of the Mexican education system, to ultimately make proposals to the government.

Mexicanos Primero's strength and growth as a civil society organization to improve public education are situated in the historical alliance between *the Sindicato Nacional de Trabajadores de la Educación* (SNTE), the National Teacher's Union; and the *Partido Revolucionario Institucional* (PRI), the Institutional Revolutionary Party. Through this alliance the teachers and their union became fundamental to the consolidation of political power by the PRI over the 70 years of its regime as the party in power. As a result, the power of *the Secretaría de Educación Pública,* the Mexican Ministry of Education (SEP), to enact education policies and to control the educational agenda throughout these years was severely limited.

Mexicanos Primero's *Relationship With the Government and the Teacher's Union*

As a first step, *Mexicanos Primero's* established a small team and began to conduct applied research on possible education policies for Mexico. The team presented a plan to the incoming president of Mexico, Felipe Calderón, from the PAN, who was to serve from 2006 to 2012. This work laid the foundation for the organization's strategy, which included influencing governmental decision makers to support the agenda and policy recommendations based on specialized advice of *Mexicanos Primero's* comparative applied research expertise. The executive director said:

> More than going along with the policy, our work was to propose. Calderón did not have in mind to transform Article 3 of the Constitution (on the governance of Public Education). We started with the basics. We had to investigate what was working, what was done elsewhere.

It is difficult to attribute the policy changes that were enacted to *Mexicanos Primero's* specialized expertise because different forces shape policy change, and causal relations are hard to determine (Brown 2008). Nevertheless, one of *Mexicanos Primero's* main goals was met at the policy level when President Calderón signed a decree that reformed Articles 3 and 31 of the Mexican Constitution in February 2012 to include the right to a free high school education. This constitutional reform made high school compulsory and ensured 12 years of schooling for all Mexican students.

In 2012, when President Enrique Peña Nieto, from the PRI, was elected and then took office, *Mexicanos Primero* continued to serve as a government ally in the creation and implementation of educational reforms. Many times, however, the CSO was a difficult ally since its constant criticism about policy implementation was not welcomed. The new president put education at the top of his agenda in his first address to the nation, on December 1st, 2012 (*Programa de Acompañamiento Ciudadano*, PAC, 2013). Before that, on October 3rd, 2012, *Mexicanos Primero* continued its research activities, developing an extensive program, *Metas 2012 to 2024* (Goals 2012 to 2024), for the new president. At the same time, *Mexicanos Primero's* team increasingly added policy-monitoring components to its repertoire of activities, assessing and overseeing education policy outcomes and implementation. Since then, the organization has been an active watchdog tracking the implementation of education policies, targeting overall educational reform, teacher evaluation and professional development, and promoting a reduction of SNTE's influence and leadership in public education in Mexico. In addition, as described in the sections below, *Mexicanos Primero* gradually refined its initial strategy of change by adding new components, including a legal approach and social advocacy work.

The primary goal of the SNTE in the years after its establishment in the 1940s was to protect the labor rights of educators. Teachers from the public and private sectors, educators with permanent and temporary positions, retired educators, and school system administrators and staff members, were all members of the union. Starting in the late 1960s, the SNTE gained increasing power over the education system. Responsibility for the administration of basic education was given to the SNTE as a political favor, and since then, the higher echelons of the everyday administration of Mexican public education has been controlled by the SNTE.

The SNTE was a powerful lobby during national political elections, and the PRI gained important votes in exchange for the promise of economic benefits for teachers. The Teacher's Union's leadership secured PRI representatives with votes in exchange for economic and political benefits. During the same period, an opposition faction within the Teacher's Union emerged: the *Coordinadora Nacional de Trabajadores de la Educación* (CNTE), the National Coordination of Education Workers. This dissident faction of the union, created in 1979, advanced an alternative political agenda to the SNTE, based on a platform that questioned the dependency and political collaboration between SNTE and PRI leadership. Throughout the years following its creation, CNTE had limited influence in national politics, given that its core membership base comprised teachers from the less developed states of Chiapas, Oaxaca, Guerrero, and Michoacán. The SNTE represents 90 percent of Mexican teachers, while CNTE has a much smaller representation. In recent years, CNTE has become more radical in opposition to educational reform, especially on the issue of teacher evaluation. Typically, the SNTE and CNTE have been at loggerheads in political struggles. From its inception, the goal of *Mexicanos Primero* was to bring a third interest group to the table, civil society, to engage in the political agenda on education reform regarding the right to education for all children in Mexico. According to Vernor Muñoz Villalobos, the United Nations' Special Rapporteur on the right to education, the Mexican education system

> reveals a profound complexity in different settings, characterized by the combination of federal and state obligations, the process of decentralization, and mostly by the atypical symbiosis between the SNTE and SEP. This symbiosis has a historical explanation, but from the viewpoint of the obligations surrounding the right to education, the State, through the Secretary of Education, is the actor of which the exercise of such right is demanded. Therefore, the mixture SNTE-SEP reveals a reciprocal subordination of atypical functions in each of the parts, which adds great complexity to the educational landscape.
> (Oficina del Alto Comisionado de las Naciones Unidas para los Derechos Humanos en México 2010, n/p)

Mexicanos Primero's Organizational Structure

Mexicanos Primero satisfies Salamon and Anheier's (1992) structural-operational definition of a CSO because the organization is non-governmental, self-governing, does not distribute profits to its board members, and is a voluntary organization. *Mexicanos Primero's* organizational structure includes a board of directors, including six prominent corporate leaders. Having started with a staff of six professionals in 2005, the organization currently has 26 staff members working in various programs. Although it is a nonprofit organization, *Mexicanos Primero* is not registered in the Registry of Civil Society Organizations, a federal registry created in 2004 to encourage the activities of CSOs in Mexico (International Center for Nonprofit Law 2012). The organization's leaders believed that including *Mexicanos Primero* in such a registry would not reflect the nature of the organization since it is not a philanthropic organization providing financial assistance and other services to vulnerable populations. Instead, *Mexicanos Primero's* aim is to advocate for education and to conduct research, generating evidence to improve education policies and quality.

As such, *Mexicanos Primero* is registered as a research organization in CONACYT (*Consejo Nacional de Ciencia y Tecnología*), the office of the Mexican government in charge of promoting scientific and technological activities. Most of the organizations registered in CONACYT are large public and private universities with a main mission of education and research activities. As a research center, *Mexicanos Primero* is required to submit an annual report about its research objectives and achievements, as well as its financial information to maintain its registration and governmental recognition as a research organization.

Donations to *Mexicanos Primero* are tax deductible. A small portion of CSOs in Mexico offer the option of a tax-deductible contribution to their donors. This option is regulated under Articles 27, 79, and 95 of the *Servicio de Administración Tributaria* (SAT) (Income Tax), which subjects organizations to specific legal and public scrutiny, incentivizing responsible fundraising for both the organization and its supporters. SAT closely monitors all donations received by *Mexicanos Primero*. The financial statements of the organization are externally audited and have been positively evaluated by PricewaterhouseCoopers S.C. since 2008.

Mexicanos Primero depends on individual donations for its operations. Prominent business leaders participate in the organization's *patronato* (board of directors) and executive committee, which gathers frequently to revise the organization's activities, finances, and campaigns. The president and board are in constant communication with the executive committee, which approves the most important strategic decisions of the organization.

Mexicanos Primero's Focus on Advocacy

Latin America is one the most starkly unequal regions in the world. While many of its countries have progressed toward improving access to free and compulsory education, inequality across socioeconomic groups, regions, and racial and ethnic groups persists. Indeed, Latin America's marginalized groups clearly lack equal educational opportunities. They receive education of a lower quality, have higher dropout rates, and fewer years of schooling. Within this context, more CSOs in the region are advocating for change with a vision to improving education conditions, including quality and access for marginalized groups.

Advocacy as a Concept and an Activity

Advocacy, from the Latin phrase, *advocare*, means "to speak to" or to argue for a particular position (Clark 2010). In his review of the literature, Clark (2010) distinguishes between two types of advocacy organizations. While some organizations engage in advocacy to advance their causes by seeking policy changes in specific policy fields, for example, in education, other associations exist essentially to provide philanthropy or educational services. In this chapter, we are concerned with the first type of advocacy organizations. These organizations, as in the case of *Mexicanos Primero*, seek to change policies and redress the underrepresented rights that political systems or economic structures exclude (Jenkins 2006), such as poor and Indigenous groups that receive an education of a lower quality in the specific case of Mexico.

We conceptualize advocacy as an organizational strategy that CSOs can adopt to fulfill their missions, where strategies are defined as the specific functions that organizations intend to fulfill (Young et al. 1999.) and are mirrored through their repertoire of activities. Advocacy strategies differ from more traditional approaches to civic and voluntary involvement, which do not seek to produce any changes in policy or institutional structures (Minkoff 1999). Building on Jenkins' (2006) seminal work on advocacy and Minkoff's (1999, 2002) studies on social change and social movement organizations, CSOs can engage in two main types of advocacy strategies. Advocacy strategies based on outsider tactics, such as protest groups, represent a direct institutional threat because groups that undertake such strategies do not accept traditional institutional channels to solve conflicts (McAdam 1996). Advocacy strategies based on reformative tactics conform to accepted institutional channels. These seek policy reform and therefore only represent intermediate forms of institutional challenges (Minkoff 1999).

Reformative advocacy strategies may include social and political advocacy activities (Jenkins 2006). Through social advocacy, CSOs attempt to influence public opinion, encourage civic and political participation, and influence the practices of private organizations (such as other CSOs), thereby shaping their organizational agendas or practices. Within the field of education, examples of advocacy include CSOs' development of education campaigns to raise awareness about specific education issues. CSOs can also get involved in capacity-building activities, such as the provision of trainings to other civil society actors or individuals to strengthen their capacity to engage in claims making.

Through political advocacy, CSOs target governmental decision makers along the various stages of policy formation—agenda setting, policy enactment, implementation, and monitoring (Jenkins 2006). Education reform is an example. For instance, in the policy enactment and implementation stages, CSOs may lobby both legislators and policy makers in the executive branch by providing specialized expertise and advice on program delivery and implementation. In the monitoring policy stage, CSOs can act as watchdogs, assessing and overseeing policy outcomes. Finally, organizations can also seek to enforce the justifiability of education rights by "taking the government to court to denounce its lack of fulfillment of its legal obligation of providing/ guaranteeing education to all its citizens" (Verger and Novelli 2012, 163).

In this chapter, we use the term "strategic advocacy framework" to depict the repertoire of activities and components through which the organization fulfills its advocacy strategy. This strategic advocacy framework differentiates *Mexicanos Primero* from traditional philanthropic education CSOs that do not seek to generate policy changes in education.

Mexicanos Primero's *Advocacy Activities*

Mexicanos Primero's advocacy strategy comprises the repertoire of all the components and activities that the organization deploys to carry out its mission. From its inception, *Mexicanos Primero's* strategy includes a strong emphasis on political advocacy, which makes it unique among civic and voluntary organizations. While the organization started by lobbying decision makers and providing them with specialized expertise about education policy, the current strategy of *Mexicanos Primero* also includes social advocacy work and a legal approach. Thus, *Mexicanos Primero* broadened its strategy by encouraging citizenship participation to improve public accountability and transparency of the education system in Mexico, and creating a spin-off organization (*Aprender Primero*) that litigates in the courts to enforce education rights. We argue that *Mexicanos Primero's* effectiveness depends on how well the organization combines *all* of its activities and components into a unified and coherent approach, sustained by a robust organizational capacity, ample financial resources, and access to decision makers.

The main activities and components of *Mexicanos Primero's* strategy that correspond with their political advocacy, social advocacy, and legal work include the following: First, the CSO has a clear policy agenda that explains the goals that it proposes for the Mexican system of education, as well as the paths through which the organization intends to reach these targets. A second component is a research team that is able to produce and publish high-quality research more quickly than many academic centers. Through evidence-based research projects (see Table 1.1), *Mexicanos Primero* seeks to provide data on the challenges of the Mexican education system and the policy solutions that it proposes to solve such problems.

Third, the organization's applied research feeds its political and social advocacy work. Political advocacy includes lobbying education public decision makers at the legislative and executive government levels and providing them with specialized expertise on education policy. Social advocacy strategies include multimedia campaigning to raise awareness on specific policy issues that mobilize participants to support the organization's demands and petitions. Strategies also include the multimedia presentation of its research findings and policy proposals to journalists and educational researchers, and using mass media to denounce abuses of power and to inform and encourage citizens' participation through online petitions. This element is supported through coordinated actions, starting with a multimedia event held in downtown Mexico City trumpeting their research findings, the publication of documents, paid advertising in the press, and provision of research information to journalists so that they can write editorials and specialized articles on the state of education in Mexico.

A fourth component that also aligns with *Mexicanos Primero's* social advocacy comprises trainings for teachers and principals and an annual award for teachers. It also includes collaborative reflection and dialogue processes that aim to connect parents and teachers. These two groups are education stakeholders that *Mexicanos Primero* believes are detached from one another, yet whose joint participation is necessary to encourage mutual responsibility among all civil society actors in education.

A fifth and final component includes an innovative legal approach that has never been used before for education in Mexico. *Mexicanos Primero* is the first CSO in the country to seek to enforce education rights and to conduct strategic litigation as a key tactic to block abuses of power from the Ministry of Education, the National Teacher's Union, and the dissident faction of the Teacher's Union.

As shown in Figure 1.1, *Mexicanos Primero's* advocacy strategy includes all five of the components and activities described above. Interdependent and implemented simultaneously, the components aim to contribute to the effectiveness of the overall framework. These components, including a continuous process of reflection that adjusts the organization's actions as needed, and access to financial resources and to decision makers, support the effectiveness of *Mexicanos Primero's* strategy. These elements are described in more detail below.

1. *A Carefully Designed Impact Policy Agenda.* According to *Mexicanos Primero*, specific changes in education policy will ensure achievement of a fairer education system. The first two focus on recovering the role and

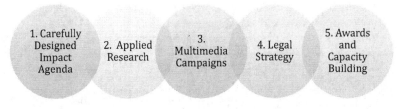

Figure 1.1 Mexicanos Primero Strategic Cycle

authority of the Mexican state in public education, which entails diminishing the power of the national teachers' union, SNTE, and professionalizing the teaching profession, so that excellent teachers and school leaders are promoted accordingly. The next two changes include improving the transparency and efficiency in the use of public resources for education, since inefficiencies ultimately undermine Mexican students' rights to a quality education. The last change involves increasing school autonomy and participation so that schools not only have more resources but also a say about their pedagogical models and curricula.

In its report *Ahora es cuando: Metas 2012–2024* (*The Time Is Now: Goals for 2012–2024*), released in 2012, *Mexicanos Primero* outlined two main elements for improving public education in Mexico: completion of schooling for all Mexican children and improved academic teaching and learning during schooling. For *Mexicanos Primero*, students should have at least 12 years of schooling, with each school year having an average of 200 days and 6.5 daily hours of instruction. This goal was met at least in policy when in 2012 President Calderón signed a decree making high school compulsory for Mexican students. Moreover, *Mexicanos Primero* also stated that Mexico needs to achieve better results in Program for International Student Assessment (PISA) tests for reading and mathematics, hoping that the country's results improve sufficiently for them to be comparable to the average of countries measured by the Organization for Economic Cooperation and Development (OECD).

In December 2012, the incoming president Peña Nieto and the leaders of the main political parties signed an agreement to promote an educational reform, supported by *Mexicanos Primero*: the guarantee that all Mexicans attain a high-quality education. Subsequent legislation was passed in 2013. In addition, a constitutional mandate created a National System of Evaluation, and strengthened the autonomous power of the National Institute for the Evaluation of Education (INEE). The Institute, created in 2001, became an autonomous entity with its own budget, charged with coordinating evaluations of education quality and teacher performance assessments.

The evaluations proposed by the reform brought radical changes to the teaching profession. The official national teachers' union, SNTE, as well as the dissident faction of the Teacher's Union, CNTE, rejected the reform, arguing that it denied teachers' fundamental rights, and accusing reformists of proposing privatization of the system. Teachers strongly rejected the teacher evaluations proposed by the law, which they perceived as punitive, and resisted through protests and strikes during school days and evaluation days. According to the reform package, teachers would now have to compete for jobs based on their skills and experience, and teachers' performance would become the main criterion for promotions, salary increases, and job termination. The exams used by the Ministry to evaluate the teachers would also allow the Ministry to strengthen in-service teacher education. Furthermore, to prevent teacher absenteeism, the reform established that teachers absent from work more than three times per month without proper justification and proof, such as a medical certificate, would be dismissed.

One aim of the reform is to decrease the number of SNTE's *comisionados sindicales*, or union delegates. A union delegate is a person paid as an in-service teacher over the course of many years, but who has set foot in a classroom only infrequently, is on leave from the classroom and school, and is devoted to administrative and organizing work for the union. During their period of service to the union, *comisionados sindicales* do not lose their tenure or seniority, keep receiving their monthly salary and benefits, and secure their labor rights guaranteed by Article 5 of the Mexican Constitution. In 2013, there were 22,353 *comisionados sindicales*, and their compensation was estimated to be 1,700 million Mexican Pesos (about US$ 102,708,000). One of the most visible campaigns of *Mexicanos Primero* has been constant challenges of the legality of the Ministry of Education's provision of wages to teachers while they are on leave for the SNTE's political work. The campaign questions the large numbers of such employees and their absence from the schools, which severely compromises the quality of education that children receive.

2. *Applied Research.* Conducting applied research to provide solutions to education problems in Mexico is the second component of *Mexicanos Primero's* political and social advocacy strategy. Through the years, *Mexicanos Primero* has built a research team dedicated to this strategy. Their research projects provide empirical evidence about the problems of the Mexican system and the policy solutions that the organization proposes. In fact, since 2009, *Mexicanos Primero's* research team has produced multiple research reports, providing evidence and expert knowledge on some of the most important problems of Mexico's education system, and making policy proposals and recommendations to solve these challenges. Without this component, *Mexicanos Primero* would not have the specialized data and research findings to support its lobbying activities and citizenship awareness campaigns. Table 1.1 provides examples of these reports, which tackle topics such as the inequities of the Mexican education system, early childhood education, and teaching the English language in Mexico.

The research reports that have been produced each year identify *Mexicanos Primero's* annual impact agenda. All of them were written to develop an integrated policy planning approach for education, with the aim of presenting a multi-sectorial analysis by introducing knowledge and perspectives from different sectors that are needed to strengthen policies in public education. Their recommendations seek to promote actions in different ministries, such as Health and Education, and to present a multilevel analysis to explain the issues faced at all levels, from the classroom to the highest position in the Ministry of Education. One example of this strategy can be seen in the 2014 research report, *Los Invisibles* (*The Invisibles*), which describes what is being done to ensure that Mexican children from birth to age six can ultimately reach their full potential, and what goals remain. The report first approaches early childhood development through a neurological and psychological perspective, then explores the relationship between access to quality early childhood education and women's access to the labor market, as well as the role of pre-K nutrition and schooling in the social and economic development of marginalized communities. The overall objective is to describe the characteristics and challenges of early childhood education in Mexico and recommend policy actions to improve the care of young children throughout the country.

In its last report, *Prof.*, *Mexicanos Primero* analyzes the professionalization of teachers and makes recommendations to improve the teaching profession. Recent governmental actions show that *Mexicanos Primero*'s call for professionalization and evaluation of teachers is increasingly being heard. Despite CNTE's condemnation of evaluation as a punitive policy, this reform has also awakened among teachers a new interest in the possibilities of teaching as a professional path. Educational reform in Mexico does not seek to fire teachers who do not get satisfactory grades in standardized evaluation, yet it punishes those who opt out of the tests (SEP 2016). During the first teacher evaluation day, on November 14th, 2015, 97.7 percent of teachers who had been registered for evaluation sat for the exam (SEP 2015). The Ministry of Education, as well as *Mexicanos Primero*, described the day as a total success for their agendas. For the SEP, improvement of teacher evaluation marked the beginning of larger reform to improve the quality of education in Mexico through the professionalization of teachers.

3. *Multimedia Campaigns.* While at first its organizational strategy prioritized conducting research and providing specialized expertise to policy makers, currently *Mexicanos Primero* also engages in research dissemination and targets citizens' participation as central elements of its advocacy strategic framework. Multimedia campaigns are therefore a central component of the organization's social advocacy work. In addition, these campaigns support the organization's agenda-setting activities. Multimedia campaigns include the repertoire of actions through which *Mexicanos Primero* presents its research findings and policy proposals, such as providing information to specialized journalists to add to their evidence-based research on shared topics of interest. Furthermore, the campaigns also implement awareness campaigns on selected education policy issues, as well as calls to action to citizens, either through direct action or social networks.

Maintaining close ties with numerous journalists is a central element in *Mexicanos Primero*'s strategy because they are priority stakeholders for the organization's research dissemination efforts, as indicated by the communications director:

> Journalists are one of our most important stakeholders. Through our work, we can say that we have been able to become an authoritative source of information for them. At the beginning we reached out to them. Now, oftentimes, they also reach out to us for reliable, quick and specialized sources of information on education in Mexico.

One example of how *Mexicanos Primero* raises awareness about education issues that matter most to its impact agenda is the documentary,

Table 1.1 Reports From *Mexicanos Primero*

Report	Year	Mission	Link
Gasto para el aprendizaje incluyente	2016	Proposals to improve public spending in education	www.mexicanosprimero.org/images/1fotos-noticias/Cartilla-MalGasto_2016_FINAL.pdf
Índice de Cumplimiento de la Responsabilidad Educativa	2016	An assessment of the compliance of local authorities in guaranteeing the right to education in Mexico	http://mexicanosprimero.org/images/icre/ICRE_2016_E-Book.pdf
Prof. Recomendaciones sobre formación inicial y continua de los maestros en México	2015	The professionalization of teachers in Mexico	www.mexicanosprimero.org/index.php/educacion-en-mexico/como-esta-la-educacion/45-estado-de-la-educacion-en-mexico/384-prof
Sorry: El aprendizaje del inglés en México	2015	The importance of the English language today and the academic outcomes of elementary school students	www.mexicanosprimero.org/index.php/educacion-en-mexico/como-esta-la-educacion/estado-de-la-educacion-en-mexico/sorry-2015
Los invisibles	2014	Early childhood education, from birth to six years old	www.mexicanosprimero.org/index.php/educacion-en-mexico/como-esta-la-educacion/estado-de-la-educacion-en-mexico/los-invisibles-2014
10 Propuestas para un Mejor Presupuesto 2015	2014	Ten proposals to improve the public spending in education	http://mexicanosprimero.org/index.php/educacion-en-mexico/como-esta-la-educacion/otros-estudios/pisa/81-talis-2014
Índice de Desempeño Educativo Incluyente (IDEI) 2009–2013	2013	Educational performance ranking	www.mexicanosprimero.org/index.php/educacion-en-mexico/como-esta-la-educacion/idei-indice-de-desempeno-educativo-incluyente/idei-2009–2013
(Mal) Gasto	2013	Public spending in education	www.mexicanosprimero.org/index.php/educacion-en-mexico/como-esta-la-educacion/estado-de-la-educacion-en-mexico/mal-gasto-2013

Ahora es cuando. Metas 2012–2024	2012	A schedule of education policies for the years 2012 and 2024	www.mexicanosprimero.org/index.php/educacion-en-mexico/como-esta-la-educacion/estado-de-la-educacion-en-mexico/ahora-es-cuando-metas-2012–2024
Metas 2011	2011	*Goals*, The state of Mexican education in 2011	www.mexicanosprimero.org/index.php/educacion-en-mexico/como-esta-la-educacion/estado-de-la-educacion-en-mexico/metas-2011
Brechas	2010	*Education Gaps*, The state of Mexican education in 2010	www.mexicanosprimero.org/index.php/educacion-en-mexico/como-esta-la-educacion/estado-de-la-educacion-en-mexico/brechas-2010
Contra la pared	2009	*Against the Wall*, The state of Mexican education in 2009	www.mexicanosprimero.org/index.php/educacion-en-mexico/como-esta-la-educacion/estado-de-la-educacion-en-mexico/contra-la-pared-2009
Índice de Desempeño Educativo Incluyente (IDEI) 2009–2012	2009	Educational performance ranking	www.mexicanosprimero.org/index.php/educacion-en-mexico/como-esta-la-educacion/idei-indice-de-desempeno-educativo-incluyente/idei-2009–2012

Source: *Mexicanos Primero*'s Campaigns and Research Reports 2009–2015.

De Panzazo (the literal translation is "a belly flop," used colloquially to express passing an exam with the minimum possible grade). Together with Cinépolis, a chain of movie houses that held pro bono screenings, *Mexicanos Primero* produced and released this documentary to raise awareness about the state of the Mexican public education system, and its multiple failures.[1] The film shed light on the problems of Mexican public education, such as students who barely understand what they read, parents who do not participate in their children's education, and more importantly, incompetent and poorly trained teachers shielded by the powerful yet corrupt national teachers' unions. The documentary generated strong negative reactions among teaching professionals and the two main national teachers' unions, as the *Mexicanos Primero* research director explains:

> For many teachers, *De Panzazo* was too negative about teachers. Mexicanos Primero generated widespread opposition among teachers, and then the organization had to work very hard to win teachers' support because of their negative reactions to the documentary. In addition, since Mexicanos Primero is being financed in part by business people who belong to prominent corporations, many believe that Mexicanos Primero is an arm of the businesses that financially support the organization. *De Panzazo* helped to create that public image of Mexicanos Primero as representing a corporate group, since a very well-known TV host from the Televisa group introduced the documentary. In part because of *De Panzazo*, National Teacher's Union leaders tend to ask teachers never to support Mexicanos Primero's initiatives.

As explained above, in 2009 the board of directors and executive director of *Mexicanos Primero* decided to encourage citizens' participation as an alternative channel through which authorities would be pressured to make changes. The communications director explained how this strategy evolved:

> We started by researching the state of education in Mexico, and during our first years we mainly made policy proposals to the president, public authorities, and decision makers. However, gradually we also started monitoring and denouncing what we thought was wrong. Later on, it was decided that we would also seek to activate citizens' participation, for example, by inviting them to sign our petitions.

Currently, *Mexicanos Primero* has 17,000 followers who regularly sign its various petitions. For example, *Mexicanos Primero* has been one of the toughest and most critical voices in the debate on the transparency and

---

1 As shown in Table 1.1, *Mexicanos Primero* has issued four reports on the state of Mexican Education: *Cumplimiento* (Compliance), 2016; *Metas* (Goals), 2011; *Brechas* (Gaps), 2010; and *Contra la Pared* (Against the Wall), 2009.

expenses of the education budget, and the use of public resources directed to education.[2] In August 2014, *Mexicanos Primero* launched the *mismanagement-ticker (abusómetro)*, a giant billboard above the Periférico, one of the main highways crossing Mexico City from North to South. The billboard, whose purpose was to raise public awareness of the inefficiency of the public education system to the general public, showed second-by-second how much money was wasted by the SEP. *Mexicanos Primero* estimated this loss to be $1,000 Mexican Pesos (about US$ 60) per second. The *mismanagement-ticker* would show this sum in real time so that citizens could see the amount increase every day. The communications director explained that through this campaign *Mexicanos Primero* invited citizens to sign *Fin al Abuso (End to Misuse)*, a petition demanding that the government promote transparency in the resources invested in public education:

> In Mexico there are economic resources for the national teachers' unions, but not for schools. Stealing is a crime. The mismanagement-ticker intended to galvanize citizens to raise awareness on this issue, since the government is not being accountable to citizens: Where are these resources going? We asked people to sign our petition, and this is how we tried to raise the profile of this issue in the public agenda. We did not have any answers from the government, but we are satisfied with the results because we are raising awareness on this problem among citizens.

4. *Professional Awards and Capacity Building*. This component of *Mexicanos Primero's* advocacy strategic framework comprises teacher training, the ABC Annual Award that recognizes public education teacher excellence, and two new pilot programs that seek to promote dialogue to provide leadership training for principals. It also includes forums for collaborative reflection and dialogue that aim to connect parents and teachers, since *Mexicanos Primero* believes that these groups do not regularly communicate with one another, but their joint participation is necessary to advance the right to education and encourage the sharing of responsibility of all civil society actors in education.

These programs advance *Mexicanos Primero's* social advocacy because they foster civic participation in education and promote changes in the attitudes of parents and teachers, as well as encourage parents' increased participation in school communities. The programs also establish relations with stakeholders that are key to *Mexicanos Primero's* mission, but that its organizational strategy did not initially prioritize, specifically with parents

---

[2] For example, see *Gasto para el Aprendizaje Incluyente 2016* and *10 Propuestas para un Mejor Presupuesto 2015*, proposals to improve public spending in Mexico's education, as well as *Mal Gasto: Estado de la Educación en México 2013*, a report on public spending in Mexico's education.

and teachers. Finally, these actions allow *Mexicanos Primero* to build stakeholder support for its advocacy campaigns.

The first initiative, the ABC Award, has benefited 86 teachers since 2008. The award recognizes excellence in teaching in public education through a specific set of criteria, including subject matter expertise, self-assessment, continuous improvement, responsibility, creation of opportunities to enable all children to learn, invitations to parents to participate, and perseverance in seeking a quality education for all students. Award recipients receive training at the *Instituto Tecnológico y de Estudios Superiores de Monterrey* (ITESM) and the Spanish University *Deusto* in Pamplona or the Autonomous University in Madrid, Spain. The Televisa Foundation also gives teachers laptops, and the *Cámara de la Industria Editorial* (Chamber of the Editorial Industry) distributes books to them. In contrast with the teachers who resist *Mexicanos Primero*'s positions, these teachers are some of the best allies for the organization. *Mexicanos Primero*'s founder and president said about those who receive the award:

> The award confers prestige to them. These teachers know us well and so they know that by no means do we intend to privatize education in Mexico. They share our cause.

In fact, *Mexicanos Primero* believes that award recipients have been key in generating bonds of trust with teachers, and recipients usually seek collaboration with the organization for research projects and consultations.

The second initiative, *En Voz Alta* (With a Raised Voice), was created from the idea of two staff members who thought *Mexicanos Primero*'s framework lacked programs that create direct ties with parents and teachers. *En Voz Alta* provides forums for parents, teachers, and children to facilitate the exchange of ideas, perceptions, and opinions. One staff member, who is now the co-director of participation initiatives at *Mexicanos Primero*, expressed how *Mexicanos Primero* related to schools before this program was created:

> We talked a lot about schools, but we did not talk with them. Their voices were not being heard.

To create such spaces for dialogue, *En Voz Alta*'s events take place in schools as the program seeks to bring education stakeholders together around a common discussion. The executive director explained that *Mexicanos Primero* aims to know and better understand parents' and teachers' perspectives and worldviews:

> We do not want to be a strictly academic organization. We do not want to say: 'we know what parents want.' Rather, we want to understand them.

The third initiative, with the support of Cambridge University and the ITESM, is an international partnership offering short courses designed to strengthen the leadership of principals and school administrators. Through a call for applications, the organization originally selected 400 principals who will interact and take the course in different facilities of ITSEM. Subsequently, a new cohort of 800 more principals was added to this program. The principals represent all 31 Mexican states and work in elementary education. In many cases, most principals had not received previous professional training for their leadership positions.

5. *Legal Approach*. *Mexicanos Primero's* legal approach is the last component of the organization's advocacy strategic framework. In 2014, its board created a parallel organization, *Aprender Primero* (Learn First), to be responsible for its legal strategies. Legal activities comprise four actions: criminal complaints; injunctions; appeals; and *amparos* (a Mexican legal term for federal judicial protection and restitution for human rights violations by government officials), which are brought on the grounds of rights violations and for the protection of the constitutional right to an education in Mexico. The creation of this parallel organization responded to the need of *Mexicanos Primero* to maintain its tax-exempt status since, under Mexican law prior to 2016, civil society organizations were not able to litigate against the government.

Mexicanos Primero uses the law to stop what it views as abuses of power by both the Ministry of Education and the national teachers' unions in addressing the right of Mexican citizens to public education. One recent example occurred in June 2015, just before the mid-term general elections. The Secretary of Public Education, Emilio Chuayffet, suddenly decided to suspend indefinitely the mandatory teacher evaluations that are one of the main components of the 2013 Law of Education. *Mexicanos Primero* saw this decision as politically opportunistic and illegal, especially before elections, and therefore filed an appeal on the grounds that the decision of the Secretary of Public Education was unconstitutional. A few days after *Mexicanos Primero* filed the complaint, a federal judge ordered the Secretary of Public Education to reinstate the evaluations, and they were carried out as originally planned. CNTE opposed the action and blamed *Mexicanos Primero* for promoting decisions that violated its rights. As *Mexicanos Primero's* executive director expressed, the organization interpreted the court's reversal of the ministerial decision as a success and an indicator of its effectiveness:

> We succeeded in our goal. We do have the document that the judge issued. Not conforming to a court order is a serious and severe error for a public official.

Aprender Primero is using class action lawsuits on behalf of students and parents without access to quality public schools. This is the first time that such lawsuits have been used for this purpose in Mexico. *Aprender Primero* legally represents parents so that their children can effectively exercise their right to education. *Aprender Primero's* general counsel explained why using this legal procedure is important for the organizational strategy of *Mexicanos Primero*:

> While the legal instrument (class action suits) was created in Mexico with the intention of its being mostly used for environmental or consumer issues, we are using this instrument to advocate for education rights. Some people in Mexico think that the law can be negotiated, but when there is a judge who is ordering that a law be applied, it has to be followed. This is why we developed this legal strategy.

One salient use of this legal approach is to make the government and educational officials accountable for investing in the infrastructure of schools across the country, especially to improve poorly built schools and to initiate construction projects in schools with no bathrooms or inadequate classrooms. *Aprender Primero* filed a class action suit on behalf of students' parents from the state of Guerrero, who agreed to be legally represented by the legal organization. Through this class action suit, *Aprender Primero* is trying to create a legal precedent so that subsequent cases have similar effects, putting pressure on the public authorities to invest in school infrastructure. This type of education advocacy strategy is new in Mexico, and its effectiveness is yet to be seen. *Aprender Primero* filed its lawsuit in January 2015, but according to its lawyer, the case has progressed slowly. The judge in charge, who is from the state of Guerrero, had never dealt with a class action suit before, and he was unsure about the legal procedures for handling the case:

> Class actions are a challenge for us as well, since it is a new legal resource with no previous history in Mexico. We even had to explain what a class action suit was to the judge. In our first class action suit we are representing parents from the Nicolás Bravo School in Xochihuehuetlán. The school has latrines instead of bathrooms. This school has 150 students, and while we only needed 30 parents to sign the petition, 88 parents agreed to confer on us their legal representation. Of course we are paying all related costs of the legal action and we are doing this free-of-charge for parents. Otherwise this would not be possible.

Also in 2014, *Aprender Primero* presented a criminal complaint against Liberato Montenegro, the leader of SNTE in the state of Nayarit. The leader and his family have been receiving teacher salaries for years without having been active as teachers. According to *Mexicanos Primero*, Montenegro and his family illegally received 14 million Mexican pesos from 2010 to 2014 (worth about US$ 852,330 USD). *Mexicanos Primero* has been increasing public awareness about "*maestros aviadores*" (fly-by teachers), a colloquial term for the people who claim a teacher's salary, but who do not teach or have any other duties in the schools, including people who have passed away and whose relatives receive a salary in their name. In Mexico, approximately 70,000 teachers receive a salary but do not teach. The legal counsel explained:

> To solve the problem of the *aviadores* we have even had to file criminal complaints. We needed to do something else other than protesting. It's like we have the constitution in our hand. We follow the law. And if necessary, we file criminal complaints against those who think they are above the law.

The issue of teachers' failure to be present in the schools and to participate in the required evaluation has moved now from the state courts to the Supreme Court of Mexico. On October 2016, the Supreme Court issued a decision that it is constitutional for the Ministry of Education to terminate a teacher's employment if, after three opportunities, he or she does not pass the evaluation. This decision is based on the main argument that those who fail to pass the evaluation are not accredited to continue in this public function as a teacher (Alzaga 2016). The filing of this decision is adding to the momentum of education reform, reinforcing the strategy of *Mexicanos Primeros* to use legal means to challenge the political system and create constructive change in education.

An Analysis of the *Mexicanos Primero* Organizational Model

Here we summarize and substantiate our three arguments about the approach and operation of *Mexicanos Primero*.

First, we conceptualize *Mexicanos Primero* as part of a broader process by which Mexican civil society has continued to expand and diversify (Verduzco 2003). *Mexicanos Primero* is representative of this expansion in the number of CSOs involved in public education in Mexico (Verduzco and Tapia 2012). In a study of the role of CSOs in education, Verduzco and Tapia (2012) noted that the number of Mexican education CSOs had increased and that their participation in education includes the implementation of activities such as teacher training, encouragement of parents' participation in schools, implementation of projects for infrastructure improvement, and evaluation of public education programs. But Verduzco and Tapia (2012) also showed that that there are few opportunities for education CSOs to monitor education policy and to encourage policy changes in education. For the most part, they implement programs designed by the Ministry of Education. In this context, *Mexicanos Primero*'s technical capacity, financial resources, and access to decision makers become central, since these elements provide the organization with the necessary capability to encourage policy changes in the education system through its repertoire of activities. *Mexicanos Primero*'s effective organizational capacity contrasts with the ways that other studies have characterized the Mexican civil society as having weak organizational capacities (Layton 2009).

Our second argument is linked to the previous one. *Mexicanos Primero* is not a corporate foundation and does not claim to represent corporate interests. Instead, the organization defines itself as an independent and pluralistic citizen-led initiative. Yet, since the organization's leadership includes powerful Mexican corporate leaders, an explanation about corporate involvement in education provides a window through which to understand its role as a civil society organization. This new approach of corporate leaders' support for education contrasts with the traditional attitudes and conduct with respect to public education in most Latin American countries, including Mexico, where businesses and leaders have transferred funds, goods, or services to support schools, but have not sought to influence educational policy, leaving it to the state (Puryear 1997) and trade unions.

From this perspective, we argue that *Mexicanos Primero* embodies a new type of involvement in education by corporate leaders, which is distinct from traditional patterns of philanthropy. *Mexicanos Primero* seeks instead to encourage policy change in education by supporting causes that monitor and demand transparency and efficiency within the education system. By supporting *Mexicanos Primero*, corporate leaders are moving away from a traditional focus on charitable donations to help groups in need. *Mexicanos Primero* is also consistent with broader philanthropic trends in Mexico, as a new generation of grant makers seeks to participate proactively in the public sphere by giving greater emphasis to efficiency, effectiveness, and co-responsibility of social actors. This emphasis contrasts with previous generations of grant makers that relied on notions of charity and depended on the work of unpaid volunteers (Turitz and Winder 2005).

Mexicanos Primero illustrates how a group of business leaders is channeling economic resources, knowledge, and networks of influence to change education policy. Paradoxically, since *Mexicanos Primero* is an advocacy CSO, and both legitimacy and accountability are key conditions for the effectiveness of CSOs that seek to encourage policy change (Brown 2008), the same corporate leadership support that strengthens *Mexicanos Primero's* organizational capacity presents additional challenges to *Mexicanos Primero's* organizational legitimacy, which is a pre-condition to become an accepted and valid policy advocate. For example, concerns about the legitimacy of advocacy CSOs raise questions such as "Who do advocacy CSOs represent?" and "In whose name do these organizations seek to encourage policy change?" In fact, the *De Panzazo* documentary showed that the close identification of *Mexicanos Primero* with the corporate media conglomerate Televisa damaged the organization's reputation among teachers as an accepted advocate for policy change in education.

The final argument that we make is that the effectiveness of *Mexicanos Primero's* advocacy depends on several components that reinforce each other in a unified and coherent strategy, supported by the financial backing of corporate leaders, access to decision makers, and a well-managed organization. Constant adaptation and strategy adjustment also support the organization's mission, so that these components reinforce each other. *Mexicanos Primero's* strategy and components operate in ways that are consistent with the work of Verger and Novelli (2012) on the role of civil society and advocacy coalitions and the Global Campaign for Education. In their study, Verger and Novelli (2012) found that CSO networks deploy similar activities such as lobbying decision makers, implementing awareness campaigns, issuing media releases, conducting applied research, monitoring education policy, and legal action.

The case of *Mexicanos Primero* shows that CSOs are not always passive entities but rather can be strategic actors that can refine and adapt their organizational strategies to fulfill their missions. From this perspective, *Mexicanos Primero's* initial approach of conducting applied research and targeting and sharing specialized expertise with decision makers evolved into a comprehensive advocacy framework with a strong focus on research dissemination and encouragement of social participation, both symbolic participation through social networks and the launching of dialogue processes at school communities. Another part of the CSO's evolution was its implementation of a legal strategy with the aim of increasing the impact of the organization, broadening its social base, and stopping what it sees as abuses of power. As Verger and Novelli (2012) argue in their study, the effectiveness of each strategy component is context specific. In fact, they contend that legal approaches are more "appropriate in countries with a legal framework and constitution that clearly guarantees the right to education" (163). Therefore, in *Mexicanos Primero's* case, while an appeal showed that the organization's intervention was effective in impeding the suspension of teacher evaluations, the criminal complaint that *Mexicanos Primero* initiated in the state of Guerrero has yet to prove it is an effective tool to generate concrete changes in education in Mexico, rather than a symbolic action.

Conclusions

In the past two decades, the field of comparative education has increasingly been concerned with the various roles and challenges of CSOs in education. Typically, research studies have focused on the role of civil society organizations as service deliverers in education (see Rose 2009). However, more recent research has engaged in a lively intellectual discussion about how CSOs are moving beyond service delivery to focus on the root causes of education disparities through advocacy strategies (Mundy 2008, 2012; Stromquist 2008; Verger and Novelli 2012).

This chapter seeks to contribute to the latter strand of research from the perspective of the organizational strategies of education CSOs by providing an in-depth look at the dynamics and the advocacy framework of *Mexicanos Primero*, a prominent Mexican CSO that seeks to advance the right to education for all students. Through the case of *Mexicanos Primero,* we analyzed how an education CSO integrates a wide repertoire of actions to fulfill its advocacy strategy. Our study also showed that in addition to having good intentions, organizational capacity and access to decision makers and resources are central elements in the effectiveness of an organization's strategy. Without doubt, corporate leadership support sustains the organizational capacity needed for *Mexicanos Primero* to encourage policy changes in education. At the same time, the strong ties with the business sector create challenges for the legitimacy of the organization that it must navigate carefully as it seeks to advance the right of all Mexicans to a high-quality education.

References

Alzaga, Ignacio. 2016, August 11. "La Suprema Corte valida la evaluación a profesores." *Milenio.com.* www.milenio.com/politica/Valida_Suprema_Corte_evaluacion_a_profesores-SCJN_evaluacion_profesores-amparo_CNTE_0_790720957.html

Bécalos. 2016. *Resultados.* Mexico City, Mexico Bécalos. https://becalos.mx/resultados

Brown, David. 2008. *Creating Credibility: Legitimacy and Accountability for Transnational Civil Society.* Sterling, VA: Kumarian Press.

Clark, John. 2010. "Advocacy." In *The International Encyclopedia of Civil Society.* ed. Helmut Anheier and Stefan Toepler. New York: Springer.

Cohen, Jean L., and Andrew Arato. 1992. *Civil Society and Political Theory.* Cambridge: MIT Press.

International Center for Nonprofit Law. 2012. *Assessing the Impact of the Fiscal Reform Agenda for Mexican Civil Society Organizations.* www.icnl.org/research/library/files/Mexico/ICNL%20Fiscal%20Reform%20Assessment%20-%20FINAL.pdf

Jenkins, Craig. 2006. "Nonprofit Organizations and Political Advocacy." In *The Nonprofit Sector: A Research Handbook.* ed. Richard Steinberg and Walter Powell. New Haven: Yale University Press.

Layton, Michael. 2009. "A Paradoxical Generosity: Resolving the Puzzle of Community Philanthropy in Mexico." *Thematic Issues on Philanthropy and Social Innovation: Issue on Community Philanthropy* 9(1): 87–102.

McAdam, Doug. 1996. "The Framing Function of Movement Tactics: Strategic Dramaturgy in the American Civil Rights Movement." In *Comparative Perspective on Social Movements.* ed. Doug McAdam, John D. McCarthy and Mayer N. Zald. New York: Cambridge University Press

Mexicanos Primero. 2012. *Ahora es cuando. Metas 2012–2024.* Mexico City, Mexico: Mexicanos Primero. www.mexicanosprimero.org/index.php/educacion-en-mexico/como-esta-la-educacion/estado-de-la-educacion-en-mexico/ahora-es-cuando-metas-2012-2024

Mexicanos Primero. 2014. *Los Invisibles.* Mexico City, Mexico: Mexicanos Primero. www.mexicanosprimero.org/images/stories/losinvisibles/Los-Invisibles_estado-de-la-educacion-en-mexico_2014.pdf

Minkoff, Debra. 1999. "Bending with the Wind: Strategic Change and Adaptation by Women's and Racial Minority Organizations." *American Journal of Sociology* 104(6): 1666–1073.

Minkoff, Debra. 2002. "The Emergence of Hybrid Organizational Forms: Combining Identity-Based Service Provision and Political Action." *Nonprofit and Voluntary Sector Quarterly* 31(3): 377–401.
Mundy, Karen. 2008. "From NGOs to CSOs. Social Citizenship, Civil Society and Educational For All." *Current Issues in Comparative Education* 10: 32–40.
Mundy, Karen. 2012. "The Global Campaign for Education and the Realization of Education for All." In *Campaigning for "Education for All": Histories, Strategies and Outcomes of Transnational Social Movements in Education*. ed. Antoni Verger and Mario Novelli. Rotterdam: Sense Publishers.
Oficina del Alto Comisionado de las Naciones Unidas para los Derechos Humanos en México. 2010. *Relator Especial sobre el Derecho a la Educación culmina su visita a México*. www.hchr.org.mx/index.php?option=com_k2&view=item&id=206:relator-especial-sobre-el-derecho-a-la-educacion-culmina-su-visita-a-mexico&Itemid=265
Programa de Acompañamiento Ciudadano (PAC). 2013. *Reforma Educativa 2012–2013*. http://pac.ife.org.mx/debate_democratico/descargas/Reforma-Educativa-2012-2013.pdf
Puryear, Jeffrey. 1997. *Partners for Progress, Education and the Private Sector in Latin America and the Caribbean*. Washington, DC: Inter-American Dialogue.
Rose, Pauline. 2009. "NGO Provision of Basic Education: Alternative or Complementary? Service Delivery to Support Access to the Excluded?" *Compare: A Journal of Comparative and International Education* 39(2): 219–233.
Salamon, Lester M., and Helmut K. Anheier. 1992. "In Search of the Non-Profit Sector I: The Question of Definitions." *Voluntas: International Journal of Voluntary and Nonprofit Organizations* 3(2): 125–151.
Secretaría de Educación Pública (SEP). 2015. *Comunicado 367*. www.comunicacion.sep.gob.mx/index.php/comunicados/noviembre-2015/1393-comunicado-367-el-evado-nivel-de-participacion-en-la-evaluacion-de-desempeno-sep
Secretaría de Educación Pública (SEP). 2016, July 7. *Mitos sobre la Reforma Educativa*. www.gob.mx/sep/articulos/mitos-generales-sobre-la-reforma-educativa
Stromquist, Nelly. 2008. "Revisiting Transformational NGOs in the Context of Contemporary Society." *Current Issues in Comparative Education* 10: 41–45.
Turitz, Shari and David Winder. 2005. "Private Resources for Public Ends: Grantmakers in Brazil, Ecuador and Mexico." *Philanthropy and Social Change in Latin America*. ed. Cynthia Sanborn and Felipe Portocarrero. Cambridge: Harvard University.
Verduzco, Gustavo. 2003. *Organizaciones no Lucrativas: Visión de su Trayectoria en México*. México: EL Colegio de México-Cemefi.
Verduzco, María Isabel and Mónica Tapia. 2012. *Organizaciones de la Sociedad Civil: Presentes en las Escuelas, Ausentes de las Políticas Educativas*. Ciudad de México: Alternativas y Capacidades. www.alternativasycapacidades.org/sites/default/files/publicacion_file/Investigaci%C3%B3nSEP-OSCs.pdf
Verger, Antoni and Mario Novelli. eds. 2012. *Campaigning for "Education for All": Histories, Strategies and Outcomes of Transnational Social Movements in Education*. ed. Antoni Verger and Mario Novelli. Rotterdam: Sense Publishers.
Young, Dennis, Bonnie Koenig, Adil Najam, and Julie Fisher. 1999. "Strategy and Structure in Managing Global Associations." *Voluntas: International Journal of Voluntary and Nonprofit Organizations* 10(4): 323–343.

2 *Mexicanos Primero*
Efforts to Account to Parents and Teachers

Constanza Lafuente and Regina Cortina

In this chapter, we explore the accountability practices of civil society organizations (CSOs), in particular the efforts to account to parents and teachers, using the case of *Mexicanos Primero*, an organization that advocates for education reform in Mexico. Complementing Chapter 1 in this volume, where we analyzed the strategic framework of *Mexicanos Primero*, here we build on concepts of accountability in the literature to identify emerging themes and thereby contribute to the research on advocacy in the comparative education field. This organization provides a good case to study the efforts of advocacy CSOs to account to parents and teachers for two reasons. First, being accountable and transparent are two of its main values. Second, in the past three years, *Mexicanos Primero* has increasingly engaged parents and teachers in debates on issues pertaining to public education.

Researchers and practitioners alike are increasingly interested in studying how education CSOs interact with, and give voice to, teachers, parents, students, and local community-based organizations (see Mundy 2008, 2012; Archer 2010; Verger and Novelli 2012). Issues for study in the field of development include how the concept of upwards and downwards accountability affect CSO behavior: *upward* with respect to the funders; and *downward* with respect to their beneficiaries, the individuals targeted by their missions who are often powerless because they do not have the economic or political resources to influence the actions of these organizations (Bendell 2006). Holding themselves accountable to these individuals—teachers, parents, and students—leads CSOs' practices to be more democratic (Bendell 2006) and holistic (O'Dwyer and Unerman 2008). This is a critical issue for investigation in development and comparative education research because CSOs tend to invest more energy and resources to being accountable to donors than to the former groups.

Two decades ago, in their seminal article "Too Close for Comfort," Michael Edwards and David Hulme (1996) argued that the preference of donors to fund advocacy, relief, and service CSOs had compromised the ability of these organizations to establish their agendas and follow their goals independent from donors' wishes. The article spurred a debate among academics and practitioners about the role of CSOs in development, their autonomy, impact, and accountability practices. In a revised version of the article, which appeared 19 years later, Nicola et al. (2015) argued that these organizations still

> face still significant constraints and contradictions in their ability to strengthen civil society given the pressures they face to be non-political, their weak roots in society, the pressures they face to be accountable 'upward' to donors rather than 'downward' to beneficiaries, and their focus on short-term projects rather than long-term structural change.
> (709)

We argue that comparative education researchers can expand their understanding of the accountabilities of education-related advocacy organizations by additionally looking at three elements that also shape such practices: adherence to mission; perspectives of what stakeholders support mission fulfillment; and role of membership in electing the organization's authorities. The aim of our analysis, presented below, is to incorporate these factors into the examination of how an organization in the education field is accountable to parents and teachers, at a time in which its programmatic efforts encourage their engagement in debates about public education. We also look at external and internal dimensions of accountability. External dimensions comprise complying to regulators and reporting to groups in CSOs' ecosystems, whereas internal dimensions refer to adherence to values and mission (Ebrahim 2010).

We focus on two research questions: What types of accountability mechanisms and stakeholders are prioritized by an education-related advocacy CSO such as *Mexicanos Primero*? And, how do external and internal dimensions relate to the organization's accountability practices? To address these questions, we adopted a qualitative approach to our study of *Mexicanos Primero* (see Chapter 1 for a description of the organization's operations and our data sources). Data analysis included identification of patterns, such as similarities and differences in the views and opinions of the individuals we interviewed, regarding the practices of *Mexicanos Primero* and the challenges of designing and implementing accountability mechanisms.

We start by defining the term *accountability* in general with special attention to education CSOs, and the roles of membership in shaping main accountability approaches. In the following section, we discuss the central elements and dimensions of accountability, presenting key themes that shed light on this case study: internal and external dimensions of the concept, organizational views of stakeholders, and how civil society organizations balance conflicting demands and needs. We then focus on our research findings on *Mexicanos Primero*, the education organization we have been studying in depth. We provide an overview of the assessment of its activities and examine its accountability philosophy. Then, we discuss the accountability mechanisms that demonstrate that the organization prioritizes donors and internal stakeholders' involvement. Next, we analyze findings on *Mexicanos Primero's* practices that are tailored to meet the needs of its various stakeholders, focusing on families and teachers who receive too little attention. In this section, we note that efforts to incorporate teachers and parents' voices to debates about public education did not result in accountability processes customized for such groups. The last section of the chapter discusses why, despite the organization's goal of advancing the right to education of all Mexicans, such groups are not prioritized in its accountability practices. Our explanation focuses on three elements: the internal dimension of accountability, which mandates adherence to organizational mission, the prioritization of stakeholders critical to mission fulfillment, and *Mexicanos Primero's* status as a non-membership organization. The conclusion offers suggestions for comparative education researchers who are investigating the roles, strategies, and practices of education-related advocacy CSOs.

The Role of Accountability in Civil Society Organizations

Citing Jenkins' (2006) perspective on nonprofit organizations' advocacy in the United States, we define the advocacy of education CSOs' as "represent[ing] the collective interests of the general public and underrepresented groups as opposed to the interests of the well-organized powerful groups especially business, mainstream social institutions and the elite professions" (307). We further note that these CSOs fulfill their missions through campaigning, raising awareness about education rights, encouraging citizen participation in education, seeking to influence policy formulation through lobbying public authorities, monitoring policies or programs, or conducting applied research to support their positions on education policy (Verger and Novelli 2012).

As civil society organizations, including those committed to education issues, proliferate at the global level and become prominent advocates for international development and policy reform, their practices have become relevant topics of study, including in the field of comparative education. Indeed, many studies now explore the roles, processes, and impact of CSOs' advocacy (Mundy 2008, 2012; Eickelberg 2012; Verger and Novelli 2012) that seeks to catalyze economic, social, and political change processes (Lewis 2007).

Comparative education builds on concepts developed in other academic fields. One strand of research within the field of non-governmental organizations (NGOs) studies the processes and implications of CSOs' *accountabilities*, a term defined as "the means by which organizations report to a recognized authority, or authorities, and are held responsible for their actions" (Edwards and Hulme 1996, 967). The idea that organizations are bound to rationalize and justify their behavior (Fry 1995) underpins such a term. This strand of research is especially important for the study of education advocacy since CSOs demand equity and transparency from federal and state authorities in the use of public resources targeted for public education. Thus, in turn, multiple groups—donors, policy makers, members, practitioners, and to a lesser extent beneficiaries—demand that they themselves be transparent in how they use their resources and responsive to the groups affected by the problems of education. Answering to stakeholders affected by their campaigns and actions is one of the crucial challenges for advocacy in the field of education, and the strategies that organizations employ to do so differ according to their particular features.

Membership Organizations

Advocacy organizations that are membership based exist to represent and serve the interests of their members (Smith 2010), and they campaign or lobby on their behalf (see Table 2.1 for an example of membership and non-membership advocacy CSOs). Members are active in electing their leadership, and therefore accountability practices in such organizations tend to prioritize them (Ebrahim 2010). At a minimum, accountability mechanisms have to demonstrate to members that the organization's campaigns or lobbying activities serve their interests and rights.

Non-membership Organizations

As discussed in Chapter 1 of this volume, advocacy organizations such as *Mexicanos Primero* represent a distinct type of philanthropic participation in education that distinguishes it from a service-oriented organization. Additionally, in contrast to membership organizations, non-membership advocacy organizations such as *Mexicanos Primero* do not conduct elections to select their leadership (e.g., board of directors). The strength of their values and missions—such as quality education for all citizens rather than member representation—gives them their legitimacy to exist and advocate for the targets they seek to achieve (Brown and Jagadanada 2007; Peruzzotti and Smulovitz 2006). Furthermore, they do not have strictly defined beneficiaries, and since they are not service-oriented organizations, they are only internally accountable to their board of directors and staff, and to external stakeholders such as donors (Ebrahim 2010).

Table 2.1 Membership and Non-membership Advocacy CSOs

Membership CSOs	*Non-membership CSOs*
— National advocacy networks	— Foundations
— Transnational advocacy networks	— NGOs
— Federations	— Think-tanks
— Unions	— Philanthropic organizations
— Professional associations	
— Grassroots organizations	

Bendell (2006) highlights various "concerns about the unaccountability of advocacy efforts" (25) of these organizations because it is advisable to explain their approaches to the individuals or entities affected by their actions and incorporate them in the planning or execution of their work. However, doing so is often an exception rather than a rule, as seen in the examples of education organizations that prioritize stakeholders other than those most directly affected by their campaigns and advocacy activities (i.e., students and parents).

The Elements of Organizational Accountability

Accountability refers to how organizations report to both authorities and stakeholders and how they are held responsible for their outcomes (Edwards and Hulme 1996). From this perspective, accountability is linked to the concepts of both justification and transparency. In other words, CSOs must clearly justify their organizational existence and decisions and make their information easily accessible. These concepts also require conforming to government regulations and accepting the possibility of penalties for lack of compliance or transparency (Ebrahim 2010).

The Internal and External Dimensions of Accountability

Drawing upon Ebrahim (2010), we conceive accountability as having internal and external dimensions that are underpinned by relations with multiple actors and embedded in specific power configurations. Moreover, such dimensions have repercussions in CSOs' accountability practices. As seen in Figure 2.1, the internal dimension refers to a CSO's adherence to its values and mission, including how it accounts to internal actors such as its staff and board of directors. This is an issue for CSOs because they are internally accountable for their purposes and principles (Ebrahim 2010) and want to remain responsive to their missions and values (Brown and Jagadanada 2007). In a seminal article on the internal aspects of accountability, Fry (1995) argued that it is also an "intrinsic experience in daily organizational life" (1) in which members of the organization feel responsible for their own actions, values, and missions. Such a view of accountability as an internal and subjective experience contrasts with the external dimension that involves complying with objective regulations, justifying the organization's actions, and reporting to multiple stakeholders, defined as the people and groups with a stake in its outcomes, performance, or sustainability (Anheier 2005). For example, a CSO's dissemination of information about its performance informs stakeholders about the extent to which the organization reaches its goals effectively and efficiently (Brown and Jagadanada 2007).

The presence of these external and internal dimensions can create inevitable tensions in organizations. For example, when donors fund particular development or education projects, CSOs may fall into *mission creep*, moving

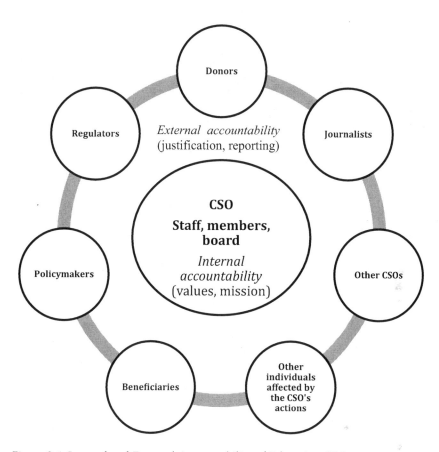

Figure 2.1 Internal and External Accountability of Education CSOs

away from their mission and expanding programs into new areas (Brown 2014) to adapt to donors' requests and preferences. Such shifting may cause internal conflicts between those who think the organization needs to adapt strategically to attract economic resources and those who prioritize internal accountability and adherence to the values and mission. Moreover, such tensions also suggest that whereas some individuals may prioritize complying with groups that provide critical resources, others may favor internal stakeholders and alignment to purpose and principles.

Restrictive and Non-restrictive Views of Stakeholders

Organizations' views of stakeholders have implications for their accountabilities, for they can have more or less restricted views of the nature of their stakeholders (Anheier 2005), with such perspectives translating into specific organizational practices related to accountability (Unerman and O'Dwyer

2006). For example, CSOs' views of stakeholders can be restricted to internal actors exclusively, such as their board of directors or staff. Organizational definitions of stakeholders can also encompass other groups or individuals that are key to mission fulfillment, such as donors, volunteers, or other CSOs that are necessary to achieve core goals. A more comprehensive view of stakeholders also includes the individuals and groups that are impacted by the actions of the organization, such as program beneficiaries or groups affected by the organization's campaigns. While inclusive understandings of stakeholders may lead to holistic organizational practices, restricted views could result in accountability mechanisms focused on fewer actors (Unerman and O'Dwyer 2006).

More restricted or limited views of stakeholders allow organizations to focus on crucial groups and individuals, avoiding the need to balance multiple demands. In their study of CSO accountability, Edwards and Fowler (2002) argue that various external and internal stakeholders (donors, policy makers, other CSOs, members, staff, volunteers, and beneficiaries) often make incompatible accountability claims, and therefore, CSOs have to continuously balance various demands. Indeed, Brown and Jagadanada (2007) contend that to ensure mission fulfillment, CSOs can be more efficient by prioritizing stakeholders that are critical for such purpose: "Since CSO[s] have many different stakeholders—donors, members, regulators, clients, allies, staffs, targets—trying to be fully accountable to all of them may be a recipe for paralysis or constant firefighting." (19).

The Accountability Component of *Mexicanos Primero*

Mexicanos Primero, which was one of the first civil society organizations in Mexico to demand accountability from federal and state authorities for the use of public resources targeted for public education, combines different activities to fulfill its mission (see Chapter 1 in this volume for a detailed description of it activities). With funding from 25 private donors, *Mexicanos Primero* has grown by implementing an increasing number of programs and gaining visibility in the discussion on education policy at the national level. Its advocacy strategies include providing expert knowledge on education to public decision makers at the legislative and executive government levels. *Mexicanos Primero* also implements campaigns to raise awareness on specific policy issues in order to mobilize citizens to support the organization's demands and petitions. The organization also has a participation program that encourages debates about public education among parents and teachers. Lastly, the organization takes an innovative legal approach, never before used for education matters in Mexico, which is based on initiating strategic litigation as a key tactic to block abuses of power by the Ministry of Education, the state-level authorities, and the national teachers' union.

Our findings show that accountability practices comprise a central activity for *Mexicanos Primero*. Staff members invest a considerable amount

Assessment and Evaluation of Mexicanos Primero's *Activities*

Consistent with the literature on organizational accountability (Ebrahim 2010), *Mexicanos Primero* accounts to its stakeholders for its finances, governance, performance in media campaigns, and program targets. It internally accounts to its board of directors and staff for how it fulfills its mission by reporting on its governance, which includes the role of the executive committee in providing strategic direction; its supervision of internal processes; and its compliance with regulations and conflicts of interest.

The organization's performance accountability includes the nature of its programs and how it implements them and meets its targets (Brown and Jagadanada 2007). *Mexicanos Primero's* activities and outputs, such as publications, are easier to measure than outcomes. The executive director underscored the challenges of evaluating the impact of its advocacy work, explaining that,

> Showing outcomes in policy decisions is important for advocacy CSOs in the field of education. These organizations do not have beneficiaries as service-oriented CSOs, and their results are less tangible. We document what we do, but it is important not to claim certain outcomes as our own. We have to pay attention to what we say publicly.

Mexicanos Primero's *Accountability Philosophy*

The factors that encourage the organization to be accountable to its stakeholders are linked to the internal and external dimensions of accountability.

Internal Factors

The individuals we interviewed for this study expressed the centrality of values and organizational culture when explaining why *Mexicanos Primero* accounts to its various stakeholders. One manager indicated that "transparency is a defining element of *Mexicanos Primero's* organizational culture, and therefore we should be coherent with our own organizational values by being transparent to ourselves and others."

External Factors

A staff member emphasized the external drivers of accountability: "Everything that *Mexicanos Primero* does is public; and therefore is publicly recorded and documented. Accountability is therefore central, since our credibility and political capital also depend on our transparency and coherence."

As this manager indicated, failing to continuously communicate with stakeholders would be a problem for the legitimacy of the organization.

The executive director pointed out that if the organization demands that the government spend public resources more efficiently, *Mexicanos Primero* also needs to be financially prudent and to continuously communicate how it spends its own resources. He used the phrase "pressing need to account" to depict the ongoing process of answering to *Mexicanos Primero's* donors. This description indeed reflects the "sense of anxiety" (O'Dwyer and Unerman 2008, 803) that external and upwards accountability generates in CSO managers. Financial resources are essential to fulfill the organization's mission, and therefore accounting to donors on a continuous basis is central for *Mexicanos Primero* because it depends on grants and contributions. In addition, because of the CSO's strategic advocacy framework, interviewees also indicated that everything that the organization does is open to public scrutiny. Otherwise, *Mexicanos Primero* would lose its relevance as an organization that seeks to advance the right to education. Being accountable increases its legitimacy. Moreover, *Mexicanos Primero* is required to regularly submit financial and legal information to government regulators to maintain tax-exempt status; as the executive director concluded, "We are a not-for-profit organization, and public oversight has become very strong lately in Mexico. All donations made to Mexican not-for profits are closely monitored."

Mexicanos Primero's Accountability Mechanisms

Mexicanos Primero's accountability mechanisms prioritize tools over participation processes and do not encourage stakeholder involvement beyond its staff and donors—many of whom are also on the board of directors—although some interviewees acknowledged that incorporating new groups would lead to improvements in its accountability practices.

Accountability Tools

Tools used by the organization to account to its stakeholders comprise financial disclosures, annual reports, research reports, and daily newsletters: concrete materials produced within a fiscal year, issued periodically, and documented (Ebrahim 2003). The annual report is one of the central accountability tools of *Mexicanos Primero*. It provides evidence of the nature and results of the multiple activities and projects that the organization implements within a fiscal year. It also makes commitments for future campaigns and actions. The financial reports provide donors, authorities, and internal stakeholders, such as the board of directors, with information about the organization's financial management to ensure them that resources are used to fulfill the organization's mission and campaigns. The bulletins are sent via email to *Mexicanos Primero's* 17,000 followers, donors, and journalists who are interested in its programs, and as such they engage followers who support *Mexicanos Primero's* various campaigns by asking them to sign online petitions. They include relevant news about Mexico's public education, interviews with *Mexicanos Primero* staff members, and information about the progress of its activities. Although all of these tools are key to providing concrete information to stakeholders, they do not encourage their participation or involvement in deciding about goals, strategies, or activities as would consultations with individuals or groups.

Accountability Processes

Mexicanos Primero's processes of accountability are less tangible and aim at generating participation, which is simultaneously their purpose and mechanism (Ebrahim 2003). Such participation includes conversations with donors around possible action plans. The literature indicates that while tools such as annual reports can be utilized in the processes of accountability, processes are more dynamic in that they seek to involve different stakeholders—donors, staff, or those affected by the CSO's actions—in multiple areas. Areas of participation could include shaping the organization's programs or strategies, participating in project implementation, collaborating in determining how to assign program resources, or monitoring programs and initiatives. Thus, the process mechanisms of accountability become essential means through which beneficiaries can interact with CSO managers and leadership, providing them with feedback on the organization's activities, goals, and strategies (Ebrahim 2003). They therefore provide more democratic (Bendell 2006) and holistic accountability (O'Dwyer and Unerman 2008). At the same time, such processes can redound in organizational learning as CSOs incorporate new perspectives to their work (Ebrahim 2010).

This organization currently has one process mechanism of accountability, which is limited to internal and upwards stakeholders, such as the board of directors, who are also its main donors. The executive committee meets with this group every six months—in March and October—to show *Mexicanos Primero's* progress and exchange ideas about possible courses of action for the organization. As the executive director explained, "Meetings are open and flexible. And this is probably one of the most important accountability instances for our organization." However, given that these bi-annual meetings are restricted to upwards and internal stakeholders, they are not examples of broader multi-stakeholder participation in the organization as described in Ebrahim's (2003) framework.

Stakeholders to Whom *Mexicanos Primero* Is Accountable

Categories of Stakeholders

Interviewees' responses emphasized *Mexicanos Primero's* external and internal dimensions of accountability by referencing stakeholders both in its wider realm of influence and within the organization itself. The president asserted:

> Mexicanos Primero is first and foremost accountable to its mission, and its cause. After that comes our executive committee and the board of directors and staff. We are also accountable to the Mexican society, because we are a tax-deductible nonprofit organization and therefore we have to be accountable to society as well.

The managers' depiction of stakeholders was more inclusive than that of *Mexicanos Primero's* president, whose view reflects the internal dimension of accountability. Managers mentioned the government, education policy makers, donors, journalists, and society as the principal stakeholders, and to a lesser extent, parents, teachers, and students. In fact, this wider view coincides with *Mexicanos Primero's* accountability practices, given that its tools are disseminated to these groups, and managers keep them continuously informed. Indeed, *Mexicanos Primero's* accountability tools reach upwards to donors, sideways to journalists, and downwards to followers and the general public. Keeping these actors updated about the progress of the organization and its activities has become more important since *Mexicanos Primero* has increasingly campaigned, petitioned to authorities, and encouraged citizenship engagement as important components of its strategic advocacy framework. Journalists are also key to influencing public opinion and disseminating the content of the reports elaborated by *Mexicanos Primero* (see Chapter 1 in this volume).

Families and Teachers as Stakeholders

Teachers and parents were not as frequently identified as prioritized stakeholders of *Mexicanos Primero's* accountability efforts, although the organization has designed and developed new programs to engage such groups in conversations on issues pertaining to public education.

Programs That Engage Parents and Teachers

In 2014, *Mexicanos Primero* created a participation program area in response to the view that autonomy should be conceded to schools and that participation of parents and teachers must be encouraged (*Informe Anual* 2014). When we asked interviewees what motivated the creation of this program area, they said that it complements the advocacy activities of the organization, because the former seeks to increase school parents and teachers' engagement in education. For example, one staff member indicated: "It is important to stop viewing policy reform from the top down and start viewing it from the bottom up, because ultimately meaningful reforms take place in classrooms." Another staff member who agreed with the former interviewee, indicated that the inclusion of new programs establishes new dialogue spaces with parents and teachers as a way of activating their engagement. The executive director added that: "We do not want to become overly academic. We do not want to say 'we know what parents want.' Rather, we want to understand their perspectives." During the interview, the executive director also expressed that incorporating teachers and parents' voices to debates about public education helps the organization to better understand their views and beliefs.

As a result of the incorporation of this new programmatic area, *Mexicanos Primero* formalized initiatives that encourage discussions among parents, teachers, and youth. *En Voz Alta* (*With a Raised Voice*) is a program created in 2013 that *Mexicanos Primero* organizes in local communities with teachers, principals, district supervisors, and pedagogical coordinators. It aims to provide a space for these groups to reflect about teaching practices, professional development, leadership, teacher evaluation, school autonomy, and challenges in public education, so that these individuals can engage in a productive dialogue without the presence of experts, policy makers, politicians, or trade unions. *Cambiemos el Rumbo* (*Changing Courses*), another program created in 2015, enables parents and grandparents to reflect together on the ways in which they can be actively involved in students' learning. Led by facilitators who are *Mexicanos Primero's* staff and volunteers, parents and grandparents cover topics such as effective learning, positive discipline, knowledge of the education system, effective participation in schools, maintaining a culture of high expectations, and the importance of daily reading for students' development. In this program, families also share their views on public education.

Nuestra Escuela Primero (*Our School Comes First*) is a program created in 2015 that fosters parents and teachers' discussions around creating a shared vision for their schools. The program incorporates the right to education as a framework for these discussions. In describing the program, one staff member indicated that facilitators ask parents and teachers "what kind of school they want, rather than the one that *Mexicanos Primero* or other organizations would like for them." He continued, "It is their right to access the school that they themselves want for their children." Through these activities, *Nuestra Escuela Primero* also aims to create a sense of belonging between schools and families, and fosters students, teachers and families' school ownership. Each school community identifies concerns about their schools and discusses possible solutions. They establish a shared mission and vision for their school and create a plan of action to implement those changes. These can include improvement plans at the school level, or engaging in discussions with local authorities to guarantee those improvements (*Nuestra Escuela Primero Report* 2016).

The programs explained previously are a shift in the organization's programmatic approach. They represent *Mexicanos Primero's* organizational learning through ten years of operation and efforts to address emergent gaps in their practices that initially did not reach school communities and did not incorporate parents and teachers' voices to discussions about public education. The organization's annual reports published between 2013 and 2016 evidence this programmatic shift. The 2013 report makes general calls for participation, and the 2014 one highlights general calls for engagement and monitoring policy makers' actions; but the 2015 report remarks the opening of spaces for dialogue and engagement in debates about public education; and the 2016 one highlights the role of social participation in achieving a better quality of education, and in guaranteeing the right to education. The annual report in 2013 indicates: "We call on social actors, in their respective areas of expertise, to take action and make concrete decisions about public policy, by responding to the question 'What can I do?'" (*Informe Anual* 2013, 7). The annual report in 2014 emphasizes the central role that citizens' play in the organization's strategy, but, it does not mention the specific functions of school communities in such conversations, highlighting instead interactions with authorities: "We call on various actors to participate in spaces of analysis, debate and claims making for education. We interact with federal and state authorities to implement and monitor education decisions" (*Informe Anual* 2014, 7). Unlike the previous report, the one published in 2015 states instead that fostering debates among students, teachers, and citizens is one of *Mexicanos Primero's* main programmatic lines of intervention. This report encourages "participation, analysis and debates about education issues with university students, teachers and citizens" (*Informe Anual* 2015, 8). Finally, the 2016 report remarks: "Social participation is a way to contribute to better learning opportunities for children and youth, and a society where more citizens are aware of their right to learn" (*Informe Anual* 2015, 26). These quotes demonstrate the increased efforts by *Mexicanos Primero* over the past four years to incorporate the voices of teachers and parents into discussions about public education.

Understanding Parents and Teachers

Our findings show that although the organization has made significant efforts to engage parents and teachers in conversations about public education, these initiatives did not translate in the participation of these groups in the organization's accountability processes. As explained previously, in addition to encouraging exchanges between managers and stakeholders, participation processes would allow stakeholders' to influence *Mexicanos Primero's* strategy, programs' design, implementation, or resource allocation. Although many teachers and parents (who are also *Mexicanos Primero's* followers) frequently receive the organization's accountability tools (e.g., reports and updates), the organization does not encourage their participation through formal accountability processes, which mainly engage donors and internal stakeholders.

Some interviewees indicated that the organization needs to improve accountability processes for parents and teachers. One of the interviewees said that accounting to these groups might be one of the challenges of the organization in the upcoming years. He expressed: "we have to create a mechanism for parents and teachers, because we want to reach these two groups." In explaining why *Mexicanos Primero*'s participation processes for parents and teachers are a developing priority, another interviewee expressed that the organization needs to understand these constituents better to be able to be fully accountable to them through customized participation processes. This perspective highlights that understanding stakeholders' views is a prerequisite to being accountable to them through participation processes. For example, when discussing the typical lack of ongoing conversations between education CSOs and teachers and parents in general, and their lack of participation in the organization's accountability processes, one manager said: "Often times, education CSOs speak about parents and teachers, sometimes even on their behalf, but don't necessarily speak to them." This manager indicated that participation programs were a first step toward understanding and creating such stronger connection with parents and teachers.

Managers interviewed for this study further highlighted additional challenges in engaging these groups through participatory processes. Said challenges included: parents' views about the right to education and different perspectives between the organization and many union-affiliated teachers on the professionalization of the teaching profession.

Parents

Some interviewees acknowledged that it is difficult for *Mexicanos Primero* to convey its message and mission to parents because often times, parents are not aware of the view that education is a right for all Mexican children, which is one of the main values of the organization. One interviewee who characterized parents as being disconnected from their schools said: "We want to reach out to these groups but it is very hard. Many parents do not know what the right to education means . . . that they are entitled to high-quality education." This interviewee explained that sharing the same view of education as a right would facilitate participation of parents in *Mexicanos Primero*'s accountability processes. Our findings show that understanding parents' beliefs and opinions about education and empowering them by promoting their engagement in their schools is a priority for the organization and according to the perspectives shared by several managers, a first step toward embracing their participation through accountability processes.

Through consultations with parents *Mexicanos Primero* confirmed a well-known fact in Mexico that low-income parents often do not think of public education as a right, but rather as a service provided by the government. In fact, in their daily speech, parents often refer to public schools as *escuelas del gobierno* ("schools of the government"), signaling that these are not spaces where their voices or input is welcome. There is a disconnect between the mission of *Mexicanos Primero* that assumes that citizens will actively engage in defending such right, and many low-income parents' beliefs and opinions about education. One interviewee indicated: "If we do not get them involved in the schools, imagine how we are going to get them to participate in *Mexicanos Primero*." This interviewee said that parents from low-income backgrounds in general do not participate in their schools, and therefore it would be hard to imagine that these groups would participate in *Mexicanos Primero's* accountability processes. He also suggested that parents' participation in school communities, through *Mexicanos Primero's* programs, precisely, seeks to boost their engagement in education.

As we explained in previous sections, using stakeholders' opinions as relevant feedback to inform decisions about programs or strategy is one of the key characteristics of participation accountability processes (Ebrahim 2003). Such processes could include creating a committee of parents and teachers to inform programs; providing these groups with representation in the organization's board of directors; or engaging them in formal discussions around programs or resource allocation to influence future programmatic decisions. Our findings suggest that the executive director expects parents and teachers' feedback to emerge from its participation programs rather than through formal accountability processes. *Mexicano's Primero's* Executive Director highlighted: "We also hope that these participation spaces will provide feedback to our organization." He emphasized that he expects to obtain rich information about parents' opinions through the dialogues and exchanges that take place with these individuals through its participation programs, highlighting that the organization can use the information obtained through those discussions as relevant feedback that managers can consider to make decisions about the organization's programs or strategy.

Teachers

As is the case with parents, the efforts to incorporate teachers' voices in debates about public education have not resulted in customized accountability processes for this group. When we asked interviwees why *Mexicanos Primero* does not engage teachers' through customized accountability practices, two staff members indicated that the organization does not have participation accountability processes in place because first its managers need to establish stronger links with these groups to be able to account to them.

Our findings indicate that teachers are another group of stakeholders who present challenges for achieving accountability. Unions' opposition to *Mexicanos Primero's* activities originate in the insistence of the organization on the need to evaluate and professionalize teachers, position that has created opposition from the two teachers' unions who have traditionally saw teachers as workers in service of the state and not as professionals charged with

Efforts to Account to Parents and Teachers 57

improving the quality of education for citizens. This has led some teachers to oppose the organization's campaigns, especially regarding teacher evaluations. For example, in 2015, an article published by the British Broadcasting Corporation (BBC) portrayed teachers' opposition to teacher evaluations—in particular members of the *Coordinadora Nacional de Trabajadores de la Educación* (CNTE), the dissident teacher union that has approximately 100.000 members—[1]with the title "Combative Teachers Who Do Not Want to Be Evaluated in Mexico." Terms such as "up in arms" (Paullier 2015) were used to depict CNTE teachers' protests against teacher evaluations, which they see as punitive and threatening. For example, one unionized CNTE member expressed that teachers oppose evaluations because these are designed "from a desk by an official who has never stepped on a school in Oaxacan territory" (Paullier 2015). The article also reported that teachers oppose evaluations for being administrative reviews intended to lay off thousands of teacher workers (Paullier 2015).

Additionally, some teachers consider that *Mexicanos Primero* demeaned them through a widely viewed documentary they produced titled *De Panzazo*, which portrays the disorderly state of Mexican public classrooms and teachers avoiding their responsibilities in an urban middle school. The executive director explained:

> There are obstacles in reaching out to teachers. . . . Many teachers—especially those affiliated with the SNTE [one of the teachers' unions with 1,619,990 members (El Diario MX 2017)]—oppose our actions, so it is very difficult to account to stakeholders that oppose our activities. Likewise, the union leaders of the dissident faction[2] refer to us as if we were politically affiliated with the Institutional Revolutionary Party (PRI), as SNTE is—and they demand teachers not to support our actions.

Although some teachers, as the ones we describe above, fervently oppose *Mexicanos Primero's* positions, others participate in the organization's programs. Teachers are not a homogeneous group. Unlike the former teachers, who resist *Mexicanos Primero's* campaigns, others—in particular those not wholly supportive of unions' positions on issues such as teacher professionalization and evaluations—participate in *Mexicanos Primero's* participation programs, including the *En Voz Alta, Nuestra Escuela Primero*, and the annual Teacher Awards (see Chapter 1, this volume, for a detailed description of this award for distinguished teachers). Rather than protesting against

---

1 CNTE was created in 1980 as a dissident faction, the National Union of Education Workers (SNTE) and is currently opposed to the educational reform passed by President Enrique Peña Nieto (Excelsior 2017).
2 The interviewee is referring to the dissident faction of the CNTE teacher union.

Mexicanos Primero's positions, these teachers choose to join spaces for dialogue around contentious issues such as professionalization, teacher evaluations, and schools' autonomy that teacher unions oppose.

When we asked the executive director about next steps to reach out to teachers, he said that *Mexicanos Primero* will continue to establish stronger links with them through their participation in existing programs and said: "We will expand the recognitions that we are giving teachers to award good teachers for their work." This finding shows that he did not mention future organizational efforts to account to them through customized participation processes. Another staff member highlighted lack of time and resources to account to all teachers. These findings suggest that although the organization plans to continue to engage teachers and to maintain dialogues with them, *Mexicanos Primero* does not intend to formally incorporate teachers' into its accountability processes in the near future because it is not central to their core goals.

Elements That Shape Mexicanos Primero's Accountability

Findings in our study illustrate that organizational efforts to incorporate teachers and parents' voices in debates about public education do not formally affect the organization's accountability processes. This is a result of *Mexicanos Primero*'s prioritization of its mission (a central aspect in the organization's internal accountability), its perspectives of which stakeholders are key to their mission fulfillment, and its status as a non-membership organization whose board of directors elects its authorities[3] explains its accountability practices. Although it seeks to advance the right to education for all Mexican citizens, the three factors we explain here help to shed light on why this organization does not fully include teachers and parents in its accountability processes.

Adherence to Mission and Values

Findings in this case suggest that while the pressing need to account to donors in part explains why *Mexicanos Primero* prioritizes them over teachers and parents, the internal dimension of accountability, in particular the role of leadership in ensuring adherence to its mission and values, is also central in shaping the organization's practices. The internal and subjective dimension of accountability highlights "felt responsibility" (Fry 1995, 1) for principles and raison d'être, and shapes organizational perceptions of stakeholders.

As demonstrated by the case findings, although some staff members and the executive director acknowledged the need to design mechanisms to enhance

---

3 In the case of non-membership organizations, boards of directors select their own members. In contrast, in membership-based organizations a general assembly of members selects the organization's authorities.

downward accountability—organizations are not uniform actors—in their daily practices, the president, executive director, and executive committee are more directly concerned with ensuring that all actions adhere to their foundational values. Thus *Mexicanos Primero's* leadership continuously assesses and refines its strategies and practices, reflecting how these help to fulfill the organization's mission to improve the quality of education through policy changes and increased transparency in the use of public resources for education.

Organizational Perspectives of Stakeholders

In the spirit of advancing its mission, *Mexicanos Primero* encourages exchanges with parents and teachers but does not give as much attention as necessary to such groups, stakeholders who do not always share its priorities. This reduces the chances of putting processes into place that elicit these groups' participation and that more closely reflect democratic models of accountability (Bendell 2006). In the case of parents, they not always conceive of education as a right. In the case of teachers that support teachers' unions, *Mexicanos Primero's* backing of the education reform and the professionalization of teachers makes them afraid that they will lose their union prerogatives. Moreover, a restricted view of its main stakeholders also helps to prevent the paralysis that could result from encouraging participation from all stakeholders in campaign or program decisions, in particular from groups that do not always share the same views regarding public education. In this way, this advocacy CSO avoids having to compromise some of its positions to accommodate new perspectives. In addition, the organization seeks active participation from its board of directors and donors, who provide *Mexicanos Primero* with guidance and resources.

Membership Status

The fact that *Mexicanos Primero* is a non-membership advocacy organization also helps explain the limited role of teachers and parents in its accountability practices. *Mexicanos Primero's* governance body, the board of directors, is not elected by a general assembly of members. While some teachers and parents may sign the petitions or participate in some of its programs, they do not elect such leaders and thus are external groups. Teachers and parents are not viewed as central stakeholders, because the organization's legitimacy depends on its mission, values, and priorities on education policy rather than on the representation of the voices that are the subjects of education rights in Mexico.

Conclusion

This case study of *Mexicanos Primero* offers a new perspective on the literature on advocacy work for education. Our analysis demonstrates some of the tensions in the accountability practices, values, and programs of a civil society organization, and provides examples of the elements that comparative education researchers can use to examine accountability practices in education advocacy. The study shows that at times, although CSOs devoted to education make efforts to incorporate parents and teachers' voices to the debates on issues that affect them, these do not redound necessarily in the increased participation of such groups in organizational accountability processes. Although the need to account to donors explains why *Mexicanos Primero's* practices prioritize some groups over others, three additional elements are important to consider. The internal dimension of accountability, which mandates adherence to organizational values, mission, and policy positions, prioritization of stakeholders critical to mission fulfillment, and *Mexicanos Primero's* status as a non-membership organization all affect the shaping of this CSO's accountability practices.

Our analysis illustrates that one of the most important dilemmas for an organization *Mexicanos Primero* is the improvement of downward accountability to teachers and parents through participation and ongoing dialogue. Felt responsibility to its mission and the role of leadership in ensuring mission fulfillment ultimately leads the organization to devote less organizational capacity to the main groups in civil society that are affected by its actions. At this point in *Mexicanos Primero's* development, it is crucial to acknowledge two resulting considerations. First, although by expanding participation processes to additional stakeholders the CSO would have to hold conversations about its strategies—in the short run, such dialogues could delay campaign or program-related decisions; in the long run, they would lead to organizational learning because ultimately, *Mexicanos Primero* could incorporate new perspectives into its work. Second, such lack of participation of teachers and parents may end up hurting *Mexicanos Primero's* legitimacy as an advocate of public education because it increases the distance between the organization and the groups in civil society that receive an education of a lower quality.

References

Anheier, Helmut K. 2005. *Nonprofit Organizations: Theory, Management, Policy.* London, UK: Routledge.

Archer, David. 2010. "The Evolution of NGO-Government Relations in Education: Action Aid, 1972–2009." *Development in Practice* 20(4/5): 611–618.

Banks, Nicola, David Hulme, and Michael Edwards. 2015. "NGOs, States, and Donors Revisited: Still Too Close for Comfort?" *World Development* 66: 707–718.

Bendell, Jem. 2006. *Debating NGO Accountability.* NGLS Development Dossier. New York and Geneva: United Nations. UN Governmental Liaison Service. www.gdrc.org/ngo/accountability/NGO_Accountability.pdf

Brown, L. David and Jagadanada. 2007 January. *Civil Society Legitimacy and Accountability: Issues and Challenge.* Working Paper No. 32. Cambridge: Harvard University, Hauser Center for Nonprofit Organizations, Civicus. www.hks.harvard.edu/content/download/68885/1248350/version/1/file/workingpaper_32.pdf

Brown, William. 2014. *Strategic Management in Nonprofit Organizations*. *Strategic Management in Nonprofit Organizations*. Burlington, MA: Jones & Bartlett.
Ebrahim, Alnoor. 2003. "Accountability in Practice: Mechanisms for NGOs." *World Development* 31(5): 813–829.
Ebrahim, Alnoor. 2010. "The Many Faces of Nonprofit Accountability." In *The Jossey-Bass Handbook of Nonprofit Leadership and Management*. 3rd ed, ed. David O. Renz. San Francisco, CA: Jossey-Bass.
Edwards, Michael and Alan Fowler. eds. 2002. *The Earthscan Reader on NGO Management*. London, UK: Earthscan.
Edwards, Michael and David Hulme. 1996. "Too Close for Comfort? The Impact of Official Aid on Nongovernmental Organizations." *World Development* 24(6): 961–973.
Eickelberg, Anja. 2012. "Framing, Fighting and Coalition Building: The Learnings and Teachings of the Brazilian Campaign for the Right to Education." In *Campaigning for "Education for All": Histories, Strategies and Outcomes of Transnational Social Movements in Education*. ed. Antoni Verger and Mario Novelli. Rotterdam. The Netherlands: Sense Publishers.
El Diario MX. 2017. "De SEP a SNTE Mil 730 Millones". May 15 http://diario.mx/Nacional/2017-05-15_a558e880/da-sep-a-snte-mil-730-millones/
Excelsior, 2017. "Coordinadora Nacional de Trabajadores de la Educación (CNTE)". www.excelsior.com.mx/topico/cnte
Fry, Roland. 1995. "Accountability in Organizational Life: Problem or Opportunity for Nonprofits?" *Nonprofit Management and Leadership* 6(2): 181–195.
Jenkins, Craig. 2006. "Nonprofit Organizations and Political Advocacy." In *The Nonprofit Sector: A Research Handbook*. ed. Richard Steinberg and Walter Powell. New Haven, CT: Yale University Press.
Lewis, David. 2007. *The Management of Non-Governmental Development Organizations*. London, UK: Routledge.
Mexicanos Primero. 2013. *Informe Anual*. Mexico City, Mexico: Mexicanos Primero www.mexicanosprimero.org/images/recursos/informes_anuales/2013/InformeAnual2013MP.pdf
Mexicanos Primero. 2014. *Informe Anual*. Mexico City, Mexico: Mexicanos Primero www.mexicanosprimero.org/images/stories/informes-anuales/InformeAnual2014.pdf
Mexicanos Primero. 2015. *Informe Anual*. Mexico City, Mexico: Mexicanos Primero www.mexicanosprimero.org/images/stories/informes-anuales/Informe_Anual_2015ch.pdf
Mexicanos Primero. 2016. *Informe Anual*. Mexico City, Mexico: Mexicanos Primero www.mexicanosprimero.org/images/stories/informes-anuales/Informe_Anual_2015ch.pdf
Mexicanos Primero. 2016. *Informe Nuestra Escuela Primero. Fase 2*. Mexico City, Mexico.
Mundy, Karen. 2008. "From NGOs to CSOs. Social Citizenship, Civil Society and Educational for All." *Current Issues in Comparative Education* 10: 32–40.
Mundy, Karen. 2012. "The Global Campaign for Education and the Realization of Education for All." In *Campaigning for "Education for All": Histories, Strategies and Outcomes of Transnational Social Movements in Education*. ed. Antoni Verger and Mario Novelli. Rotterdam: Sense Publishers.

O'Dwyer, Brendan and Jeffrey Unerman. 2008. "The Paradox of Greater NGO Accountability: A Case Study of Amnesty Ireland." *Accounting, Organizations and Society* 33(7–8): 801–824.

Paullier, Juan. June 15, 2015. "Los Combativos Maestros que no Quieren ser Evaluados", *BBC Mundo*. www.bbc.com/mundo/noticias/2015/06/150617_mexico_reforma_educativa_maestros_evaluacion_jp

Peruzzotti, Enrique and Catalina Smulovitz. eds. 2006. *Enforcing the Rule of Law: Social Accountability in the New Latin American Democracies*. Pittsburgh: University of Pittsburgh Press.

Smith, David Horton. 2010. "Grassroots Associations." In *The International Encyclopedia of Civil Society*. ed. Helmut Anheier and Stephan Toepler. New York: Springer.

Unerman, Jeffrey and O'Dwyer, Brendan. 2006. "Theorizing Accountability for NGO Advocacy." *Accounting, Auditing and Accountability Journal* 19(3): 349–376.

Verger, Antoni and Mario Novelli. eds. 2012. *Campaigning for "Education for All": Histories, Strategies and Outcomes of Transnational Social Movements in Education*. Rotterdam, The Netherlands: Sense Publishers.

3 The Student Movements to Transform the Chilean Market-Oriented Education System[1]

Cristián Bellei, Cristian Cabalin, and Víctor Orellana

One of the most important changes in the Chilean political system in recent decades was the establishment of automatic registration and voluntary voting in 2012. Its political objective was to increase youth participation in elections, which had been low since democracy was restored in 1990. The lack of electoral participation among youth was explained during the 1990s as an expression of general apathetic behavior, and young people were considered "the 'whatever' generation" (*La generación "No estoy ni ahí"*) due to their supposed apolitical attitudes and limited motivation to be involved in public affairs (Moulián 2002; Muñoz Tamayo 2011). Despite their disinterest in electoral and partisan politics, there is evidence that Chilean youth had profound criticisms of society (Duarte Quapper 2000) and a high level of interest in public and social problems, especially those related to inequity and arbitrary discrimination issues (Schulz et al. 2010).

The dictatorship of Augusto Pinochet (1973–1990) fiercely implemented a process of de-politicization and demobilization of Chilean society and contributed to apathy among youth. Moreover, in 1980, a comprehensive neoliberal reform in Chile restructured the national education system and the delivery of social services. As a result of this reorganization, social services such as health care, pensions, and education began to be delivered by market dynamics and private companies, and the public sphere was seen as increasingly less relevant, further diminishing the will for political participation. After 1990, the political process was increasingly isolated from the rest of the society, and the democratic governments that followed the dictatorship did not emphasize the creation of new spaces for social organization and participation (De la Maza 2010; PNUD 1998; Moulián 2002; Ruiz Encina 2015; Huneeus 2016).

---

[1] A previous version of this article was published by *Current Issues in International Education*. Bellei, Cristian and Cristian Cabalin. 2013. "Chilean Student Movements: Sustained Struggle to Transform a Market-oriented Educational System." *Current Issues in Comparative Education*.15(2): 108–123.

In fact, since 1990, citizen mobilizations were a minor component of the democratization process in Chile. But then, beginning in 2006, youth, specifically students, began playing a crucial role as organizers of massive demonstrations around educational demands. Although students represent only a part of youth movements, social movements related to education have historically had a powerful impact on political and socio-cultural structures (Gill and DeFronzo 2009).

Education has been the major focus for protest by Chilean youth expressing their malaise. During the last decade, Chilean society has been shaken by two sharply critical and powerful student movements,[2] one in 2006 and another five years later. Students' demands, which have been echoed in civil society and the political arena, have centered on some of the Chilean education system's structural elements. Since Chilean education has increased some basic indicators of coverage and improved student outcomes during the last decade, such as enrollment rates, one might wonder why student protests have been so intensive and sustained during this period.

In this chapter, we describe and analyze these student movements to illustrate how students can be relevant political actors in education debates. We analyze this issue, linking students' demands with the distinctive institutional features of the Chilean educational system that have made it a case study of privatization in the educational sector since the 1980s. Accordingly, first, we explain the main features of the Chilean education system, including its extreme degree of marketization, which provided the institutional context of the movements. Next, we analyze the key components and characteristics of the 2006 and 2011 student movements: We describe basic features of the two movements separately and then identify their key common elements, especially from an education policy perspective. Thus, we mainly focus on the link between students' proclamations and demands and the market-oriented institutions that prevail within the Chilean education system. Additionally, we describe and analyze the relationship between the student movements and other civil society organizations that participated in both movements, although with varying levels of coordination. Finally, to evaluate the efficacy of the student movements in the education policy arena, we identify their impact on Chilean educational debates and examine some general implications for policy making.

Overview of the Two Student Movements

During the first of the student movements, in 2006, secondary students were in the streets for more than two months participating in massive marches

---

[2] According to Tilly and Wood (2015), a social movement is a political complex that combines "(1) campaigns of collective claims on target authorities; (2) an array of claim-making performances including special-purpose associations, public meetings, media statemens, and demonstrations; and (3) public representations of the causes's worthiness, unity, numbers, and commitment" (8). As we will show, the recent Chilean student movement satisfies all of these fundamental characteristics.

and protests (Domedel and Peña y Lillo 2008). Their protest was called the "Penguin Revolution" because of the black and white uniforms worn by high school students. Struggling against the neoliberalism of the Chilean education system, demonstrators attracted strong political attention and paved the path for the next big movement, which was led by university students beginning in 2011. *The New York Times* called this second movement the Chilean Winter because it occurred around the same time as the Arab Spring revolutions against some regimes in the Middle East (Barrionuevo 2011). For seven months during 2011, university and secondary students shook the country with a movement that democratic administrations had not seen for more than 20 years. Strong popular support, charismatic leaders, and a powerful critique of educational inequalities were some of the characteristics of this movement.

Chile's Market-Oriented Education System

Chilean education in general—and specifically secondary education—has frequently been presented and is seen by many as an exemplary case in Latin America. This image, which has deep historical roots, has rested in the past two decades on the active role of governments in implementing education policies aimed to increase both quality and equity (Delannoy 2000; OECD 2004). Also, since it initiated market-oriented institutional reforms in the 1980s, Chile has become an important case study of marketization in the educational sector and has gained attention in international and comparative education policy studies (Patrinos et al. 2009; Witte 2009).

The Nature of Chile's Primary and Secondary Education System

The focus of Chile's reforms was administrative decentralization of the public education system through the creation of a voucher system for both public and private schools. Additionally, these reforms involved the implementation of universal academic achievement tests and the institution of evaluation systems and monetary incentives for teachers. Since the mid-1990s, Chile has also been very active in implementing innovative, large-scale policies on educational improvement, including the introduction of computer technology, increased school hours, curricular reform, and diverse forms of teacher education (Cox 2003; OECD 2004). Moreover, during the last two decades, enrollment in both secondary and postsecondary education has increased rapidly, and 15-year-old Chilean students significantly improved their performance on reading skills according to PISA tests between 2000 and 2009 (OECD 2011).

Thus, at first glance, it may not seem obvious why Chilean students protested so vigorously. In our opinion, the key to understanding what triggered student protests is a critical review of the way that Chilean education has been organized as a market-oriented system, and the consequences of that institutional arrangement. Although there is no single definition of what constitutes the organization of the education system as a market, reviewing the academic and education policy literature allows identification of three key elements: school choice, competition among schools, and privatization of education (Friedman 1955; Chubb and Moe 1990; Howell and Peterson 2006). In general, from a market-framed perspective, schools should compete for families' preferences, and families should have the freedom to choose a school for their children. Ideally, families should be aware of the relative quality of the available options and use such information when making their choices, in this way rewarding the best schools and forcing the worst ones to either improve or leave the market. Finally, schools need to be able to distinguish their offerings from competing schools and accommodate families' preferences; in order to do that, schools should enjoy high flexibility (i.e., few government regulations) in terms of curriculum and management. Market proponents encourage the expansion of private schools precisely because they expect them to be able to react more productively to market pressures, thereby improving both quality and efficiency in education.

Market proponents also think public schools should be radically restructured to be competitive, giving school administrators the freedom to manage schools in a business-like manner. In the education policy arena, market-oriented reformers also promote state-issued vouchers as the public funding mechanism of schools. In their view, parents should be free to use the vouchers in the schools where they prefer to educate their children. Educational vouchers therefore simultaneously promote parental choice, competition among schools, and school privatization (for reviews of empirical evidence of these proposals see Ladd 2003; Levin and Belfield 2006; Witte 2009). Although these ideas have been intensively discussed by educational experts and policy makers around the world for many years, they have been implemented in only highly restricted ways in a few countries. From a comparative perspective, Chile is one of the countries where market-oriented reforms in education have been implemented more drastically.

In Chile, since the early 1980s, the driving force for the expansion of K–12 and postsecondary education comprised the supply and demand dynamics of a market-oriented approach: Minimum requirements were set for the creation of new institutions and for the receipt of public funding, public and private institutions had to compete for families' preferences, and a universal voucher system (a state subsidy paid according to the student's monthly attendance) was established for funding private and public schools on equal terms. Also, to produce local market competition, public school administration was transferred from the national Ministry of Education to local municipalities (Gauri 1998; Bellei and Vanni 2015). Since then, from the Ministry of Education's perspective, there has thus been no difference between a public and a private subsidized school.

Although some relevant changes were introduced during the post-dictatorship period, the structural elements of the marketized system have been deepened, rather than modified (Cox 2003; Bellei and Vanni 2015). For instance, in 1993, a family fee-charging mechanism was designed, creating what is called "shared funding," a co-payment system that allowed (and encouraged) private schools and secondary public schools to charge a tuition fee without losing access to the state subsidy. As a consequence, secondary education in Chile is compulsory but not free of charge, with subsidized private primary schools also charging tuition fees. Moreover, vouchers continue to be the fundamental mechanism for financing schools, with the amount steadily increasing. In 2008, an additional voucher was created to target the poorest 30 percent of students, making them more attractive to private sector institutions. Both the co-payment system and the additional voucher provision for the poorest students were seen by Chilean policy makers as policy instruments to enhance market dynamics within the educational field.

As mentioned, a key criterion of Chilean education policies and regulations throughout this period was the establishment of "equal treatment" by the state of both for-profit and not-for-profit academic institutions. This meant that private schools had access to the same public resources, including funds for supplies, equipment, and infrastructure development (Bellei et al. 2010).

The Effects of Chile's Primary and Secondary Education Reforms

Market-oriented reforms in Chilean primary and secondary education have been evaluated in terms of their effects on both equity and quality. Although this literature is abundant and highly complex, in general, the conclusions are not positive for market proponents (Gauri 1998; Carnoy and McEwan 2000; Hsieh and Urquiola 2003; Bellei 2009; Valenzuela et al. 2013). The evidence shows that market-oriented reforms have increased educational inequities in terms of social and academic segregation, social inequality of academic achievement, and school discriminatory practices (OECD 2004). In addition, no significant gains in overall educational quality have been associated with market-oriented reforms in education.

Chile's Market-Oriented Postsecondary Education System

Market dynamics have also prevailed over the expansion of postsecondary education in Chile since 1980, but more so during the last decade when enrollment at this level really exploded (Ginsburg et al. 2003; Brunner 2009; Meller 2010). "Traditional" universities, which existed before the neoliberal reform of 1980, have had to self-fund, through mechanisms including increasingly higher tuition fees. Students whose families cannot afford the cost of higher education have access to loans that are highly subsidized by the state. Still, the growth of postsecondary enrollment mainly occurred through the creation and expansion of private institutions. These institutions do not participate in the public admission system—based on academic records and admission tests—and charge students or their families the entire cost of the education provided. Since the mid-2000s a system of state-guaranteed loans has been administered by private banks with high interest rates for students who attend these private institutions. This regulatory and policy framework has slowly evolved into a higher education market, greatly differentiated by types of institutions and highly stratified in relation to price, quality, and the social composition of the student body (Brunner and Uribe 2006; Meller 2010).

Public Reponse to Chile's Market-Oriented Education System

As we will discuss, the 2006 and 2011 student movements centered on this market-ruled education system, with students demanding a more active role of public institutions in education, especially to guarantee an acceptable standard of quality and reduce inequities. Nevertheless, it is important to note that the market-oriented education policies described above were only part of an extensive neoliberal reform implemented by Pinochet's dictatorship in the 1980s, in which privatization policies were seen as instruments to reduce state power and eliminate welfare state institutions (Moulián 2002; Cavieres 2011). In this sense, students' push against neoliberal education policies crystallized the main criticism leveled against the broader neoliberal social and economic policies in Chile: high degrees of inequality between a privileged minority and the majority of the population (Sehnbruch and Donoso 2011; Orellana 2012). In fact, according to World Bank indicators, Chile has one of the most unequal income distributions in the world (World Bank 2013).

The 2006 Secondary School Student Movement

In May 2006, thousands of students aged 15 to 18 were in the streets. They created the "Penguin Revolution," in which education became both a political and public issue (Domedel and Peña y Lillo 2008). This movement—which soon received the support of university students and teachers' union organizations—was the most significant set of demonstrations in Chile since the return of democracy in 1990. In the first stages of this movement, the demands were for provision of free transportation passes to students and the elimination of the fees associated with the university admission exam. However, the student struggle subsequently shifted to focus on the poor quality and high inequality of Chilean education in terms of attainment, resources, and opportunities. In the political arena, the students' target was the Constitutional Law of Education (LOCE), the legal foundation of the education system enacted by the Pinochet regime in 1990. The market-oriented institutions in Chilean education exist within a very complex legal framework that includes LOCE, the Chilean Constitution (also imposed by Pinochet's regime), the voucher law, and several other specialized regulations. LOCE, in particular, reduced the state to a subsidiary role in education, and promoted privatization in education; it was strongly opposed even during the dictatorship, and university students and professors had been unsuccessfully calling for its repeal since the return to democracy.

More generally, although their platform evolved over time and became manifest with diverse emphases, the students' critique consisted of four key elements: (1) the demand for free education, (2) the defense of public education, (3) the rejection of for-profit educational providers, and (4) the elimination of schools' discriminatory practices. As a whole, the ideals of this Chilean student movement represented a rejection of the rule of market dynamics in education.

First, *students demanded free education*, which implied a rejection of the co-payment system at the school level. Indeed, the fact that Chilean government-funded public and private schools are allowed to charge tuition fees to families had been a highly controversial issue, since compulsory education was considered formally "free" in Chile. International organizations, such as the United Nations International Children's Emergency Fund (UNICEF) and United Nations Educational, Scientific and Cultural Organization (UNESCO), also expressed doubts about the consistency of these practices with international treaties on the matter. Moreover, at the higher education level, Chile was the country with the highest private spending and fees in relative values among all of the member states of the Organization for Economic Cooperation and Development (OECD 2011). Thus, Chilean families—not the state—paid for the accelerated expansion of postsecondary education in recent years. The high private cost of postsecondary education was a key concern shared by both high school and university students.

Second, *students advocated for public education*. In fact, since the establishment of the market-oriented system in the early 1980s, public education at the secondary level had declined from 75 percent to less than 40 percent of the national enrollment between 1980 and 2012. Similar declines occurred at the primary level. These declines in the proportion of public education mark a reversion to the situation at the beginning of the Republic in the mid-nineteenth century. Importantly, the reduction in public education provision could not be attributed to the superior quality of private education: Evidence shows that, under equal conditions, student achievement is similar for public and voucher private schools (Bellei 2009). Students' advocacy for strengthening public education implied a demand for increased state responsibility over public education, including the creation of a funding system that gives priority to public institutions and the end of municipalized primary and secondary school administration.

Third, *students rejected for-profit private providers of education*, especially when their profits were obtained from public funds. In Chile, since the creation of the voucher system, the fastest growing sector in primary and secondary education had been that of for-profit institutions, which received state subsidies on equal terms with not-for-profit private institutions and public institutions (Elacqua 2009). Even some private universities that were required to formally constitute themselves as nonprofit institutions engaged in business strategies that circumvented the spirit of the legislation (Mönckeberg 2007), further discrediting profit-making in education in the public opinion. The pursuit of profit in education had been defended by neoliberal supporters as the engine that invigorates growth and fosters innovation. In contrast, students saw it as the source of many undesirable practices in education, including discrimination against students from low-income families and students with low academic abilities, low-quality education services, and the uncontrolled growth of low-cost undergraduate programs with low employability outcomes.

Finally, *students pushed for the elimination of discriminatory practices by schools* and the reduction of social segregation in education. Chilean schools were applying arbitrary mechanisms for selecting students, both in the admission process and throughout students' academic trajectories. Primary and secondary schools selected students based on past performance, prediction of future performance, personal behavior, family income, and other family characteristics. These selective mechanisms were especially prevalent in private institutions, including schools receiving state funding (Bellei 2009; Contreras et al. 2010). Many of these practices had long been denounced by international organizations and human rights advocates as detrimental to students' right to education. Nevertheless, Chilean political and judicial institutions defended the notion of "free enterprise" in the educational market, giving educational providers the freedom to set their own rules to admit and expel students, arguing that the mere existence of public schools was enough to guarantee the right to education (Bellei and Pérez 2000; Casas et al. 2001; Casas and Correa 2002). These selection methods, especially those that discriminated based on family income, help explain the very high levels of socioeconomic segregation in Chilean schools (Valenzuela et al. 2013), which placed Chile's education system as one of the most socioeconomically segregated of all the countries participating in the Program for International Student Assessment (PISA; OECD 2010).

Traditionally, Chilean student protests and movements had been of two types. The first was clearly motivated by general political issues, a sort of student-level expression of the national political process. In such cases, the basic dynamics were those of the political parties that were either against the government or defending it, and whose agendas were framed in the context of social transformation and ideological struggle. The other type of student protests evolved around claiming concrete demands and direct benefits for students, and thus expressed a clear interest group perspective. In these cases, the list of demands emphasized benefits that the students, as stakeholders, hoped to get from authorities at the national or institutional level.

The movement led by secondary students in 2006 certainly had clear aspects of both traditions: It articulated a solid ideology on educational issues, and it brought a significant list of concrete demands to the negotiating table. Importantly, according to public opinion surveys, the movement generated strong and widespread support and sympathy from the majority of Chile's citizens. In our view, this happened for two reasons. First, the student movement managed to formulate a demand for equal opportunity in education around the idea of the right to quality education. Second, it identified specific foundations of the market-oriented framework of Chilean education that had to be dismantled to accomplish that goal. In other words, for these twenty-first-century citizens, access to the school system was not enough. For them, equitable access to quality educational content and processes was the essential criterion to apply when evaluating whether or not the right to education has been guaranteed.

The 2011 Higher Education Student Movement

On April 28, 2011, 8,000 university students marched in different cities across Chile. The following month, a second march doubled that size. These two protests were only the beginning of what would become one of Chile's major historical student movements, producing an array of demonstrations that had enormous citizen support. The protests lasted for seven months, during which university students, united by the Chilean Student Confederation (*Confederación de Estudiantes de Chile, Confech*), organized 26 weekly massive marches or public demonstrations (some involving around 100,000 people in Santiago), took over their universities, held assemblies, and changed the public agenda in education. Camila Vallejo, president of the Student Federation of the University of Chile (FECH) and Giorgio Jackson, president of the Student Federation of Catholic University (FEUC) were two of the most charismatic leaders of the movement. With the help of leaders of regional universities, they transformed the protests into a national movement that attracted international attention. Soon, students from nontraditional private universities and secondary students joined and actively participated in the demonstrations (Cabalin 2012; Salinas and Fraser 2012).

Initially, students demanded more resources for public education and free access to universities for poor and middle-class students. Subsequently, however, they called for free postsecondary education for all, arguing that the state must guarantee the right to education from early childhood to higher education with equal conditions for all social classes. President Sebastián Piñera's administration rejected the demand for free education. However, in an attempt to placate the protesters, the government created new university scholarships to support students from the lowest socioeconomic quintiles. This action did not appease protesters, however, because their main concern was the high cost of tuition and other fees associated with loans incurred by the majority of students. In Chile, families financed 73 percent of higher education costs, a figure that greatly exceeded the 16 percent average for OECD countries (OECD 2011). Funding their education was a major issue for students because tuition and fees at Chilean universities were some of the most expensive among OECD countries in relative terms (OECD 2011).

The university system reforms in the 1980s created the conditions for the proliferation of new private universities in the following years, many of them of very low quality. In addition, according to available indirect evidence, some of these universities, despite being legally defined as nonprofit organizations, yielded substantial financial returns for their owners, thanks to legal subterfuge. Students criticized the for-profit spirit in the higher education system. Moreover, their discourse reflected notions of social justice in education by rejecting the subsidiary role of the state in education, promoting universal non-discriminatory access to free education, and requesting progressive tax reform to publicly fund education (Vallejo 2012; Figueroa 2013; Jackson 2013).

To accomplish their goals, students implemented a comprehensive political strategy, extending their collaborative networks and involving additional stakeholders, such as the teachers' union, workers' unions of various labor sectors, and several civil society organizations. Student organizations and some of the leaders of the movement published brief policy documents and disseminated information extensively through traditional and new media. Through these actions, they contributed to the re-politicization of public discussion about education and social equality issues. The political strategy of the movement allowed for the integration of different social demands in a national movement for education (Lustig et al. 2012).

After months of public demonstrations, students became more than protesters in the streets: They became political actors with a clear agenda of transformation and a coherent discourse about justice in education (Cabalin 2012). Consequently, leaders of the student movement were recognized by policy makers as relevant players in the education policy debate. For example, the Minister of Education negotiated directly with these leaders to create a first set of policies to answer their demands; then, the Chilean Congress invited them to discuss the 2012 National Budget Law.

Ideologically, the university student movement criticized the neoliberal system of education, as the Penguin Revolution had done five years before (Orellana 2012). Students asked for structural changes, such as a stronger state role in regulating and controlling educational institutions, a new system of public funding for education, reinforcement of the public universities in terms of funding and expansion, and the effective exclusion of for-profit organizations as educational providers at all levels. All these issues became part of the education policy debate in Chile, and both the Government and the Parliament have discussed different proposals to tackle them.

Characteristics of the Chilean Student Movements: A New Generation of Activists

It is possible to say that Chilean students are part of a new generation of political actors in education. From a sociological perspective, Chile is experiencing a transition from a passive generation to an active one. Karl

Mannheim (1952) argued that traumatic experiences play a key role in the production of a generational consciousness. For Chilean adults and policy makers, Pinochet's dictatorship was that kind of traumatic episode. Consequently, they assumed as common sense the economic and political institutions implemented by the military regime and only partially modified by the democratic administrations. Nevertheless, students who marched in 2006 and 2011, most of whom were born in the era of new democracy, were not part of that story: They felt free to question the limits defined by the previous generation.

Edmunds and Turner (2005) offer a valuable explanation for understanding the shift from a passive generation to an active one. For them, this change occurs when a generation is "able to exploit resources (political/educational/economic) to innovate in cultural, intellectual or political spheres" (562). They conclude that a new generation is created when young people combine these resources and innovations with political opportunities and strategic leadership. Looking at the student movements from this perspective, Chile is experiencing the birth of a new generation. In order to deepen this idea, we identified four features that characterize the recent Chilean student movements (Bellei et al. 2014): (1) persistence; (2) the combination of short-term and more structural, long-term demands; (3) innovative forms of organization and communication; and (4) multiple mechanisms of coordination.

The first element that stands out regarding the movements has been its persistence. In effect, the first series of massive protests took place in 2001 and was known as the *mochilazo* (demonstration with backpacks). The *mochilazo* was articulated around a demand for better conditions and lower pricing of public transportation for students, and also a greater presence of the state in terms of regulating and controlling the service. A high level of support among students in Santiago got the government to consent to their demands after a complex negotiation process. The *mochilazo* not only broke the public silence of students in a post-dictatorship context, but it also showed the emergence of new forms of student organization. These forms involved a combination of the traditional student council (strengthened by the organizational and participation policies of the mid-1990s) with less structured but strongly coordinated and highly motivating student assemblies. The *mochilazo* experience also made clear that government institutions did not know how to process these demands and that the traditional form of political negotiation was not effective in this new scenario. Some of these key features of the *mochilazo* were direct antecedents of the 2006 and 2011 student movements, which continued with less intensity during 2012 and 2013. Student organizations involved in those processes have been accumulating knowledge and refining their political action in the field for a decade.

A second feature of the student movements was the ability to articulate not only short-term demands (e.g., lower costs of transportation, higher quality school equipment and infrastructure), but also a set of demands that aimed to transform structural aspects of the education system. For instance, the students challenged the regulatory legacy of the Constitutional Law of Education, which was enacted on the very last day of the Pinochet government in 1990. They also protested against privatization, tuition charges, and discriminatory practices in the selection of students. The Penguin Revolution of 2006 made clear that the discourse of protest and claims of the student movement was becoming increasingly stronger and more systemic, going well beyond a simple list of student demands.

A third element characterizing the student movements was innovation in the ways that students organized and expressed themselves. Because of Chilean young people's general mistrust of traditional forms of political delegation and representation, students tried alternative ways to organize politically. To be clear, political militancy and traditional forms of student action did not disappear, but they were complemented, and in many cases surpassed, by new forms of participation, representation, and decision-making processes. For instance, in organizational terms, students used diverse types of assemblies and designated spokespeople with horizontal practices for deliberations and decision making. For instance, when these organizations communicated to influence public opinion, student leaders acted more like assembly "spokespeople" than authority figures representing an organization. In both 2006 and 2011, student organizations also implemented sophisticated mass media communication strategies, guided by leaders with outstanding and refined communication skills.

Finally, the coordination process also changed, mainly through the intensive use of new communication technologies and instant messaging. These technologies allowed students to summon a group quickly, widely, and cheaply, and also to spread their ideas and protest outcomes through the mass media. Indeed, the media did not replace but rather complemented the creation of various face-to-face initiatives, which gathered representatives based on geographic (e.g., Santiago areas) or institutional (e.g., vocational secondary schools) criteria. Forms of public demonstration were also diverse. This is particularly noticeable when looking at the 2011 student movement, during which students employed numerous forms of pressure to influence authorities and also adopted a different range of strategies to spread their message to the general public. Strategies included traditional marches, strikes, and occupations, but also new forms of public demonstration, such as massive dances, carnivals, street debates, and videos and performances in public places.

Collaboration Between the Chilean Student Movements and Civil Society Organizations

The Chilean student movement has had a complex relationship with its society. While trying to maintain a high level of autonomy from political parties, the movement has evolved, developing an increasingly close relationship with other social actors. Schematically, we have identified three types of collaboration between the student movement and civil society organizations, which correspond to social, political, and technical realms.

Coordination With Other Social Actors

The various examples of the student movement's coordination with other social actors constitute the movement's most relevant relationships with the rest of the society.

On one hand, it is noteworthy that the expansion of the student movement itself was possible because there were changes in traditional student organizations. Historically, the main student organizations were anchored to the traditional institutions (essentially public, and some nonprofit private, particularly at the university level). Nevertheless, precisely because of the market reforms, enrollment increased, mostly linked to the creation of new (mainly for-profit) private institutions. As a consequence, the student organizations' incorporation of students attending these new institutions became a critical challenge for the movement.

Interestingly, high school students exerted the leadership in this endeavor (Thielemann 2016). In 2001, the Coordinator Assembly of Secondary Students (ACES) historically led by the traditional and most prestigious public high schools, incorporated a massive number of students from subsidized private schools; then, in 2006, students from non-subsidized private schools actively participated in the Penguin Revolution. Student leaders regard the expansion of the social bases of the movement—to the point of including students from high socioeconomic status—as a partial explanation for the strength of the 2006 movement (Ruiz Encina 2013). At the university level, the process was somewhat slower. In 2005, students attending nontraditional private universities created their own organization (Student Confederation of Private Higher Education, CONFESUP); it was not until 2011, the year of the largest student demonstrations, that students from nontraditional private universities began to formally participate in CONFECH, a process of convergence that ultimately implied the dissolution of CONFESUP (Figueroa 2013 Jackson 2013).

On the other hand, the Chilean student movement has also actively worked to create a broader social movement for changes in education, collaborating with actors from inside and outside the education system. During 2006, secondary students led the formation of the "Social Bloc," an instance of coordination among many relevant organizations of education stakeholders, including the teachers' union, non-professional educational assistants, parents' organizations, and university students (Domedel and Peña y Lillo 2008). The 2006 Social Bloc attained a high level of organization and presence as a grassroots movement, and developed a programmatic common position on key issues of education policy; the strength of these groups acting together was then very important in the social dialogue organized by the government to process the students' demands (Bloque Social 2006; Garretón et al. 2011).

The 2011 student movement did not replicate the Social Bloc experience; nevertheless, they articulated a close dialogue with the presidents of the traditional universities, a collaboration that still remains (Jackson 2013). Also in 2011, the student movement started to frequently collaborate with social movements outside education (mainly supporting and participating in street demonstrations together), such as the national union federation—Central Units of Workers (CUT), the movement against the private Pension Fund Managers (AFP), and the environmental movement, among others.

Relationship With the Political Parties

As explained, the Chilean student movements have been markedly focused on political issues and consequently have participated in the political arena as a relevant player. Certainly, their relationships with the political parties (especially with the two dominant political coalitions that have governed during this period) have been tense, since the students' demands imply a significant departure from the current situation.

However, the student movement has increasingly intensified its links with political actors. In 2001, the secondary students were completely autonomous from the political parties, a feature frequently highlighted by leaders and analysts (Salazar 2012; Ruiz Encina 2013), and which reflects the wide gap mentioned between the traditional politics and the young generation after the dictatorship (Thielemann 2016). Nevertheless, this situation changed in the following years. Some of the most notable leaders of the 2006 Penguin Revolution were in fact members of traditional political parties, a circumstance that also occurred in the 2011 movement. Additionally, some leaders of the university student movement were then elected as popular representatives to the Chilean Parliament in 2014 (either from the Communist Party or as independent members of new left-wing movements). Moreover, from the student movement have emerged some new left-wing political parties and organizations that are currently creating a political alliance to fully participate in the national elections.

Certainly, we can also identify a convergent tendency from the political field, since some of the leaders of traditional political parties in turn supported the student movement. For example, despite the fact that the Socialist Party was a member of the government coalition in 2006, many of its leaders and even some congressmen joined the student movement (Ruiz Encina 2007; Domedel and Peña y Lillo 2008). But the closest relationship between the student movement and the traditional political parties materialized during the 2011 movement, when virtually all the congressmen outside the right-wing government coalition supported the students' demands and collaborated with the student leaders to advocate for institutional and regulatory changes in education (Concertación 2011). In fact, when Michelle Bachelet was once again elected president in 2014, leading a center-left political coalition, her political agenda in education was strongly based on the demands of the student movement.

Collaboration With the Academic Field

Finally, the student movements have also benefited from the technical support provided by academia, including advice, research, workshops, and training. For example, the secondary student movement was advised by the human rights NGO *Servicio de Paz y Justicia*, SERPAJ (Service of Peace and Justice) in 2001 and the research program Observatorio Chileno de Políticas Educativas, OPECH (Chilean Observatory of Education Policy) of the University of Chile in 2006 (OPECH 2009). The university student organization CONFECH has received technical assistance since 2005 from the independent academic center *Centro de Estudios Nacionales de Desarrollo Alternativo*, CENDA (National Center for Research on Alternative Development) and TERRAM, a nonprofit foundation focused on alternative development and environmental issues (TERRAM and CONFECH 2005; CONFECH 2011). This type of collaboration has been mostly restricted to periods of negotiation with the authorities, when student leaders have to present their ideas during the legislative process, or to produce programmatic documents.

Overall, it is possible to affirm that the relationship between the student movement and civil society organizations has increased and solidified over time. Actually, while the movement had a kind of episodic existence before 2011, acting almost exclusively around large demonstrations and other forms of protests, since 2011, the student movement has maintained a constant presence as a social actor in the public arena. With this purpose, as shown, the movement developed different types of collaboration with congressmen, social organizations, academic centers and political parties, and some of its leaders have also became political leaders who actively participate in the public policy debate.

Conclusion: Chilean Student Movement and Education Policy

The student movement is an ongoing process and some demands are still being subjected to political debate, but there has already been a tremendous impact on Chilean education policy (Bellei et al. 2008; Bellei et al. 2010; Bellei et al. 2014). The fact that a student movement strongly affected both the policy debate and policy decisions represents a significant change for Chilean society and is of major interest from a comparative perspective on education policy.

In fact, after the secondary student protests in 2006, all changes seemed possible. President Michelle Bachelet created an Advisory Presidential

Council for Quality in Education to debate and propose policy guidelines for improving both quality and equity in education (Garretón et al. 2011). After six months of deliberations, the Advisory Council presented a report that encompassed a wide variety of recommendations, including strengthening the right to access quality education free of charge; holding the state responsible for guaranteeing quality education; establishing a new public Quality Assurance Agency in education; reforming the institutional system of public school administration; and significantly modifying the current funding system (Consejo Asesor Presidencial 2006).

President Bachelet embraced some of the Advisory Council's recommendations and proposed a "new architecture of Chilean education." She sent to Parliament an ambitious set of legal reforms, including a new General Law of Education that replaced the previously described Constitutional Law of Education, the creation of a Regulatory Agency in Education to control the legal aspects of the system, the creation of an Agency for Quality in Education, and the reform of the administration of the public schools. Each of these reforms, except the last, was approved. In our view, the combination of a sense of emergency and social pressure from the student movement, with the consensus view generated by the Advisory Council, gave policy makers a new perspective, opened unexpected political opportunities, and resulted in a policy agenda focused on institutional transformation of the Chilean education system. Interestingly, when Bachelet was once again elected president in 2013, she proposed an agenda even closer to the students' demands: to end public funding to for-profit private schools, to prohibit discriminatory student selection by the schools, including academic selection; to eliminate the co-payment, making all subsidized schools free; and to strengthen public education, replacing the municipalities by new public institutions specialized in administrating public schools. Bachelet's Government framed these proposals as a way to reduce the marketization of the Chilean school system, changing the current paradigm by transforming education into a social right, a rhetoric that clearly resembled that of the student movement.

The impact on higher education of the 2011 student movement has also been considerable (Bellei et al. 2014). President Piñera and his Ministers of Education did disagree with some of the most emblematic demands of the students, including free education, giving priority to public education, and ending public funding to for-profit providers; however, the administration implemented a new system of public funding that increased the proportion of students with higher education scholarships and significantly reduced student loan interest rates. That administration also passed a tax reform measure to fund new education policies and proposed a major change in the accreditation system of postsecondary educational institutions, which is currently being discussed by the Chilean Parliament. Further, the Parliament created special commissions to investigate some private universities regarding potentially illegal for-profit activities (see Mönckeberg 2007;

Commission Report 2012). Finally, the education policy issues raised by the student movement were intensively debated in the 2013 presidential campaign in Chile, and the new Government proposed a higher education reform that attempted to answer some of the key students' demands (including to gradually make postsecondary education free of tuition), which is currently being debated in the Chilean Parliament.

In general terms, students framed their struggle within the "politics of meanings" in education, which refers to the conceptual dispute about the role of education in society (Simons et al. 2009). Thus, from an education policy perspective, the student movements challenged public understanding of the education system because the students rejected the notion of the problem-solving approach supported by traditional policy makers (Garretón et al. 2012). Certainly, students participated in defining problems in education, but they also participated in the discussion of policy implications, roles that played with high levels of sophistication supported by their different forms of collaboration with other organizations, including some from the academic field. Thus, policy research oriented CSOs played relevant roles by strenghtening students' capacities to contribute to these interpretations of education problems in Chile. As political actors in the educational arena, students tried to be part of the contexts of influence, text production, and practice (Bowe et al. 1992). These aspects of student participation extended the notion of the policy cycle beyond the diagnostic–design–implementation–evaluation cycle that characterizes the bureaucratic structure and technocratic process of education policy creation (Reimers and McGinn 1997). The student movements not only highlighted new problems, but also new interpretations of those new problems. Such interpretations implied the need for systemic changes in education, which were outside the framework of reference for Chilean policy makers.

From this perspective, the consequences of the student movements are also evident beyond the educational field. Certainly, the collaboration of the student movement with other civil society organizations and social actors contributed to this end. The debate about education in Chile has been linked to larger social concerns, such as Chile's unequal income distribution and the country's lack of participatory institutional structures, two issues that ultimately question the legacy of the dictatorship in both the economic and the political fields. Thus, as a social movement, the activist students can be considered "agents actively engaged in the production and maintenance of meaning for constituents, antagonists, and bystanders" (Benford and Snow 2000, 163).

During the last decades, the design and evaluation of public policies in health, poverty, and education increasingly became technical activities mainly engaged in by professional experts. Consequently, students—like the beneficiaries of social programs—were traditionally excluded from the processes of engaging education policies. The Chilean student movements showed the limits of this practice. Increasingly, policy makers, especially in matters like education, now need to consider social and cultural aspects to design and evaluate policies; introducing participatory processes into the policy cycle seems to be the most appropriate way to accomplish this (Reimers and McGinn 1997; Rizvi and Lingard 2010).

The shift toward increased participation by local actors in the education policy process proceeds in a direction opposite to that of the international organizations in the education policy field, whose growing relevance is well documented. In fact, education policies have become enmeshed with the new dynamics of globalization, where the main concern is to increase economic competitiveness. Within this context of "policy borrowing," supranational organizations—such as the World Bank and other regional banks, International Monetary Fund, UNESCO, and OECD—have created a network of interactions with public authorities, policy making agencies, and transnational corporations that highly influence national education policies (Ball and Youdell 2007). This was also the case for Chilean higher education in the last decades (Ginsburg et al. 2003). Nevertheless, since public policies can also express a collective will to solve social problems, the 2006 and 2011 student movements reminded Chilean policy makers that despite a globalized policy field, they are still socially and locally accountable.

References

Ball, Stephen J., and Deborah Youdell. 2007. *Hidden Privatisation in Public Education*. London, UK: Education International.

Barrionuevo, Alexei (2011, August 5). "With Kiss-Ins and Dances, Young Chileans Push for Reform." *The New York Times*. www.nytimes.com/2011/08/05/world/americas/05chile.html

Bellei, Cristián. 2009. "The Private-Public School Controversy: The Case of Chile." In *School Choice International*. ed. Paul E. Peterson and Rajashri Chakrabarti. Cambridge, MA: MIT Press.

Bellei, Cristián, Cristian Cabalin and Víctor Orellana. 2014. "The 2011 Chilean Student Movement Against Neoliberal Educational Policies". *Studies in Higher Education*, 39(3): 426–440.

Bellei, Cristián, Daniel Contreras and Juan Pablo Valenzuela. 2008. "Debate sobre la educación chilena y propuestas de cambio." In *La agenda pendiente en educación. Profesores, administradores y recursos*. ed. Cristián Bellei, Daniel Contreras and Juan Pablo Valenzuela. Santiago de Chile, Chile: Ocholibros Ediciones.

Bellei, Cristián, Daniel Contreras and Juan Pablo Valenzuela. 2010. "Viejos dilemas y nuevas propuestas en la política educacional chilena." In *Ecos de la revolución pingüina* ed. Cristián Bellei, Daniel Contreras and Juan Pablo Valenzuela. Santiago de Chile, Chile: Editorial Pehuén.

Bellei, Cristián, Pablo González and Juan Pablo Valenzuela. 2010. "Fortalecer la educación pública, un desafío de interés nacional." In *Ecos de la revolución pingüina*. ed. Cristián Bellei, Daniel Contreras and Juan Pablo Valenzuela. Santiago de Chile, Chile: Universidad de Chile-UNICEF.

Bellei, Cristián and Luz María Pérez. eds. 2000. *Ciclo de Debates: Desafíos de la Política Educacional: Tensión entre derecho a la educación y libertad de enseñanza*. Santiago de Chile, Chile: UNICEF Chile.

Bellei, Cristián and Xavier Vanni. 2015. "The Evolution of Educational Policy in Chile, 1980–2014." In *Education in South America*. ed. Simon Schwartzman. New York: Bloomsbury.
Benford, Robert D. and David A. Snow. 2000. "Framing Processes and Social Movements." *Annual Review of Sociology* 26: 611–639.
Bloque Social. 2006. *La crisis educativa en Chile, propuesta al debate ciudadano*. Santiago, Chile: Bloque Social.
Bowe, Richard, Stephen J. Ball and Anne Gold. 1992. *Reforming Education and Changing Schools. Case Studies in Policy Sociology*. London, UK: Routledge.
Brunner, José Joaquín. 2009. *Educación superior en Chile. Instituciones, mercados, y políticas gubernamentales (1967–2007)*. Santiago de Chile, Chile: Ediciones Universidad Diego Portales.
Brunner, José Joaquín and Daniel Uribe. 2006. *Mercados universitarios: el nuevo escenario de la educación superior en Chile*. Santiago de Chile, Chile: Ediciones Universidad Diego Portales.
Cabalin, Cristian. 2012. "Neoliberal Education and Student Movements in Chile: Inequalities and Malaise." *Policy Futures in Education* 10(2): 219–228.
Carnoy, Martin and Patrick J. McEwan. 2000. "The Effectiveness and Efficiency of Private Schools in Chile's Voucher System." *Educational Evaluation and Policy Analysis* 22(3): 213–239.
Casas, Lidia and Jorge Correa. 2002. "Conductas discriminatorias, abusivas e infundadas en contra de estudiantes en la selección y marginación en los establecimientos de educación básica y media: Diagnóstico y caracterización del problema." *Revista de Derechos del Niño* 1: 173–223.
Casas, Lidia, Jorge Correa and K. Wilhelm. 2001. "Descripción y análisis jurídico acerca del derecho a la educación y la discriminación." *Cuadernos de Análisis Jurídico* 12: 157–230.
Cavieres, Eduardo A. 2011. "The Class and Culture-Based Exclusion of the Chilean Neoliberal Educational Reform." *Educational Studies* 47(2): 111–132.
Chubb, John and Terry Moe. 1990. *Politics, Markets, and America's Schools*. Washington, DC: Brookings Institution Press.
Commission Report. 2012. *Informe sobre el funcionamiento de la educación superior*. Valparaíso, Chile: Commission Report, Cámara de Diputados de Chile.
Concertación de Partidos por la Democracia. 2011. "Respuesta de los partidos de la Concertación de Partidos por la Democracia a la propuesta de los actores sociales Bases para un Acuerdo Social por la Educación." *Declaración Pública*. Valparaíso, Chile: Concertación.
CONFECH. 2011. *Bases técnicas para un sistema gratuito de educación*. Santiago, Chile: CONFECH.
Consejo Asesor Presidencial. 2006. *Informe final para la calidad de la educación*. Santiago de Chile, Chile: Consejo Asesor Presidencial.
Contreras, Dante, Paulina Sepúlveda and Sebastián Bustos. 2010. "When Schools Are the Ones that Choose: The Effects of Screening in Chile." *Social Science Quarterly* 91: 1349–1368.
Cox, Cristián. 2003. Las políticas educacionales de Chile en las últimas dos décadas del siglo XX. In *Políticas educacionales en el cambio de siglo*. ed. Cristián Cox. Santiago de Chile, Chile: Universitaria.
De la Maza, Gonzalo. 2010. "La disputa por la participación en la elitista democracia Chilena" [Special Issue]. *Latin American Research Review* 45: 274–297.

Delannoy, Francoise. 2000. "Education Reforms in Chile, 1980–98. A Lesson in Pragmatism". *Country Studies, Education Reform and Management Publication Series*. Washington, DC: Education Reform and Management Team, Human Development Network-Education, The World Bank.

Domedel, Andrea and Macarena Peña y Lillo. 2008. *El mayo de los pingüinos*. Santiago de Chile, Chile: Ediciones Radio Universidad de Chile.

Duarte Quapper, Klaudio. 2000. "Juventud o juventudes? Acerca de cómo mirar y remirar a las juventudes en nuestro continente." *Última Década* 13: 59–77.

Edmunds, June and Bryan S. Turner. 2005. "Global Generations: Social Change in the Twentieth Century." *The British Journal of Sociology* 56(4): 559–577.

Elacqua, Gregory. 2009. "For-Profit Schooling and the Politics of Education Reform in Chile: When Ideology Trumps Evidence." *Documento de Trabajo n 5, Centro de Políticas Comparadas de Educación*. Santiago de Chile, Chile: Universidad Diego Portales.

Figueroa, Francisco. 2013. *Llegamos para quedarnos. Crónicas de la Revuelta Estudiantil*. Santiago de Chile, Chile: LOM Ediciones.

Friedman, Milton. 1955. "The Role of Government in Education." *Economics and the Public Interest* 2(2): 85–107.

Garretón, Manuel Antonio, María Angélica Cruz and Félix Aguirre. 2012. "The Experience of Presidential Advisory Committees in Chile and the Construction of Public Problems". *Revista mexicana de sociología* 74(2): 303–340.

Garretón, Manuel Antonio, María Angélica Cruz, Félix Aguirre, Naim Bro, Elías Farías, Pierina Ferreti and Tamara Ramos. 2011. "Movimiento social, nuevas formas de hacer política y enclaves autoritarios. Los debates del Consejo Asesor para la Educación en el gobierno de Michelle Bachelet en Chile". *Polis* (Santiago, Chile) 10(30): 117–140.

Gauri, Varun. 1998. *School Choice in Chile. Two Decades of Educational Reform*. Pittsburgh: University of Pittsburgh Press.

Gill, Jungyun and James DeFronzo. 2009. "A Comparative Framework for the Analysis of International Student Movements." *Social Movement Studies* 8(3): 203–224.

Ginsburg, Mark, Oscar Espinoza, Simona Popa and Mayumi Terano. 2003. "Privatisation, Domestic Marketisation and International Commercialisation of Higher Education: Vulnerabilities and Opportunities for Chile and Romania within the Framework of WTO/GATS." *Globalisation, Societies and Education* 1(3): 413–445.

Howell, William G., and Paul E. Peterson. 2006. *The Education Gap: Vouchers and Urban Schools* (rev. ed.). Washington, DC: Brookings Institution Press.

Hsieh, Chang-Tai and Miguel Urquiola. 2003. "When Schools Compete, How Do They Compete? An Assessment of Chile's Nationwide School Voucher Program." NBER Working Paper Series. Cambridge, MA: National Bureau of Economic Research.

Huneeus, Carlos. 2016. *La democracia semisoberana: Chile después de Pinochet*. Santiago, Chile: Taurus.

Jackson, Giorgio. 2013. *El país que soñamos*. Santiago de Chile, Chile: Debate.

Ladd, Helen. 2003. "Introduction." In *Choosing Choice: School Choice in International Perspective*. eds. David Nathan Plank and Gary Sykes. New York: Teachers College Press.

Levin, Henry and Clive Belfield. 2006. "The Marketplace in Education." In *Education, Globalization & Social Change*. eds. Phillip Brown, Hugh Lauder, Jo-Anne Dillabough, and A. H. Halsey. London, UK: Oxford University Press.

Lustig, Nora, Alejandra Mizala and Eduardo Silva. 2012. "¡Basta ya! Chilean students say 'enough'." In *The Occupy Handbook*. eds. Janet Byrne and Robin Wells. New York: Back Bay Books.
Mannheim, Karl. 1952. "The Problem of Generation." In *Essays in the Sociology of Knowledge*. ed. Karl Mannheim. New York: Oxford University Press.
Meller, Patricio. 2010. *Carreras universitarias: rentabilidad, selectividad, discriminación*. Santiago de Chile, Chile: Centro de Investigación Avanzada en Educación and Uqbar Editores.
Mönckeberg, María Olivia. 2007. *El negocio de las universidades en Chile*. Santiago de Chile, Chile: Debate.
Moulián, Tomás. 2002. *Chile actual: anatomía de un mito*. Santiago de Chile, Chile: LOM Ediciones.
Muñoz Tamayo, Víctor. 2011. "Juventud y política en Chile: Hacia un enfoque generacional." *Última Década* 35: 113–141.
OECD. 2004. *Reviews of National Policies for Education. Chile*. Paris, France: OECD.
OECD. 2010. *PISA 2009 Results: Overcoming Social Background. Equity in Learning Opportunities and Outcomes*. Paris, France: OECD.
OECD. 2011. *Education at a Glance 2011*. Paris, France: OECD.
OPECH. 2009. *De actores secundarios a estudiantes protagonistas*. Santiago, Chile: OPECH, Observatorio Chileno de Políticas Educativas.
Orellana, Víctor. 2012. "Sobre el malestar social con la educación y la energía del movimiento social: el primer paso del siglo XXI." In *Es la educación, estúpido*. ed. Matías del Río. Santiago de Chile, Chile: Editorial Planeta.
Patrinos, Harry Anthony, Felipe Barrera-Osorio and Juliana Guáqueta. 2009. *The Role and Impact of Public-Private Partnerships in Education*. Washington, DC: World Bank.
PNUD. 1998. *Desarrollo humano en Chile—1998. Las paradojas de la modernización*. Santiago, Chile: PNUD.
Reimers, Fernando and Noel F. McGinn. 1997. *Informed Dialogue: Using Research to Shape Education Policy around the World*. Westport, CT: Praeger.
Rizvi, Fazal and Bob Lingard. 2010. *Globalizing Education Policy*. New York: Routledge.
Ruiz Encina, Carlos. 2007. "Qué hay detrás del malestar con la educación?" *Análisis Del Año*, 2006: 33–72.
Ruiz Encina, Carlos. 2013. *Conflicto social en el neoliberalismo avanzado. Análisis de clase de la revuelta estudiantil en Chile*. Buenos Aires, Argentina: CLACSO.
Ruiz Encina, Carlos. 2015. *De nuevo la sociedad*. Santiago, Chile: LOM-Fundación Nodo XXI.
Salazar, Gabriel. 2012. *Movimientos sociales en Chile*. Santiago, Chile: Uqbar Editores.
Salinas, Daniel and Pablo Fraser. 2012. "Educational Opportunity and Contentious Politics: The 2011 Chilean Student Movement." *Berkeley Review of Education* (1). http://escholarship.org/uc/item/60g9j416
Schulz, Wolfram, John Ainley, Julian Fraillon, David Kerr and Bruno Losito. 2010. *ICCS 2009 International Report: Civic Knowledge, Attitudes, and Engagement Among Lower-Secondary Students in 38 Countries*. Amsterdam, The Netherlands: IEA.
Sehnbruch, Kirsten and Sofia Donoso. 2011. "Chilean Winter of Discontent: Are Protests Here to Stay?" *Open Democracy*. www.opendemocracy.net/kirsten-sehnbruch-sofia-donoso/chilean-winter-of-discontent-are-protests-here-to-stay

Simons, Maarten, Mark Olssen and Michael A. Peters eds. 2009. *Re-Reading Education Policies. A Handbook Studying the Policy Agenda of the 21st Century.* Boston: Sense Publishers.

TERRAM and CONFECH. 2005. "Repensar la educación superior. Un nuevo sistema de acreditación socio-económica". *Análisis de Políticas Públicas*, No. 31.

Thielemann, Luis. 2016. *La anomalía social de la transición. Movimiento estudiantil e izquierda universitaria en el Chile de los noventa (1987–2000).* Santiago, Chile: Tiempo Robado.

Tilly, Charles and Lesley J. Wood. 2015. *Social Movements 1768–2012.* 3rd ed. New York: Routledge.

Valenzuela, Juan Pablo, Cristián Bellei and Danae De los Ríos. 2013. "Socioeconomic School Segregation in a Market-Oriented Educational System: The Case of Chile." *Journal of Education Policy* 29(2): 217–241.

Vallejo, Camila. 2012. *Podemos cambiar el mundo.* Santiago de Chile, Chile: Ocean Sur.

Witte, John F. 2009. "Vouchers." In *Handbook of Education Policy Research.* eds. Gary Sykes, Barbara Schneider. and David Plank. Washington, DC: AERA.

World Bank. 2013. *GINI index (World Bank estimate).* Accessed March 2, 2017. http://data.worldbank.org/indicator/SI.POV.GINI

4 The *Círculos de Aprendizaje* Program in Colombia
The Scaling-Up Process

Laura María Vega-Chaparro

This chapter analyzes the scaling-up of an education innovation at the national policy level. The selected case is the *Círculos de Aprendizaje* (Learning Circles) program, which is based on the *Escuela Nueva* model (New School), an educational approach well known both in Colombia and internationally. *Escuela Nueva* is a cost-effective educational model that offers quality primary education in rural multigrade schools. Schools operated under this model receive specific curricula, teacher training, guidance on community, and administrative strategies to transform the school, and they emphasize intercative instruction, flexible promotion, and personalized learning processes (Gaviria in press, Colbert et al. 1993, Schiefelbein 1993).

The novelty of *Círculos de Aprendizaje* is that it reaches marginalized populations, especially internally displaced children who have been out of the educational system for at least six months, with the ultimate goal reintegrating them into the educational system. *Círculos de Aprendizaje* is an educational innovation originally designed by *Fundación Escuela Nueva Volvamos a la Gente* (FEN), a Colombian non-governmental organization (NGO), and scaled up nationally by the Ministry of Education because of the positive results demonstrated in the pilot phase of the program.

Often times, NGOs aim to affect change in public education by scaling up their programs and mainstreaming them into governmental programs. *Círculos de Aprendizaje* is a good example of how an education innovation designed by a local NGO influences education policy, although it was not an easy or straightforward process (Wils 1995, Edwards and Hulme 1992). Thus, the study described in this chapter seeks to identify the most relevant lessons from scaling-up the *Círculos de Aprendizaje* by analyzing the mainstreaming process and its implementation at scale. The chapter highlights the conditions under which education programs can be implemented at scale, with a focus on sustainability, local ownership, and quality interventions that effectively change practices in schools and classrooms. It also illustrates the complexities of the roles of Colombia's Ministry of Education and FEN, the NGO that originally created the innovation, and how the participation of FEN as a technical advisor during the scaling-up process could improve the quality and sustainability of the initiative. In fact, based on the experience of scaling up *Círculos de Aprendizaje*, the NGO has been making some changes in its advocacy and program implementation strategies.

This case is an example of political advocacy (Jenkins 2006) that influences education policy at the national level in Colombia. The chapter first briefly describes the methodology of the study. It next provides information about FEN, the NGO that designed the *Círculos de Aprendizaje* program, and the program itself, including the complex and multilayered structure created to scale it up. The chapter then explores the challenges faced during implementation of program nationally, focusing on the work of the individuals involved at all levels, defined priorities, and efforts at local contextualization. The last section presents the conclusions of the study and recommendations for further research.

Study Methodology

The analysis in this chapter is based on a study that explored the way that *Círculos de Aprendizaje* responded to the education needs of marginalized students (see Vega-Chaparro 2014). The study was designed as qualitative vertical case study (Bartlett and Vavrus 2009) that addressed different levels of education governance, while contrasting different regional cases. Using maximum variation criteria (Maxwell 2005), the study considered four different regions in Colombia where the program was being implemented in 2012.

Data, collected between June 2012 and May 2013, included document analysis, classroom observations, and interviews. I conducted 51 semi-structured interviews and observed 22 *Círculos de Aprendizaje* classrooms (see the Table 4.1 for information about the interviews and observations). Document analysis considered a wide range of documents regarding the program and public policy for marginalized populations; a total of 15 documents were analyzed in depth.

The study reviewed the participation of two organizations that acted as program operators and implemented the program in the regions. Each organization was in charge of implementing *Círculos de Aprendizaje* in two regions included in this research.

Validity tests used in this research included the collection of rich data, triangulation of data collection methods, validation by participants, comparison of different sites, and long-term involvement with the participants.

All the quotes presented in this chapter are drawn from interviews or transcripts of field notes and translated by me.

Fundación Escuela Nueva Volvamos a la Gente

Fundación Escuela Nueva Volvamos a la Gente (FEN) is an education organization founded in 1987 by Vicky Colbert, one of the creators of

Table 4.1 Description of Conducted Interviews and Observations

Data Collection Method	Number
Interviews*	
Ministry of Education Authorities	5
Program Operators	4
Regional Teams	22
Secretariats of Education	3
Auditing Firm	1
Education Institutions (Mother Schools)	14
Fundación Escuela Nueva Volvamos a la Gente	2
Classroom Observations**	
Region 1	8
Region 2	2
Region 3	6
Region 4	6

* Approximately two hours each. The description of the interviewees appears in the next section of this chapter.
** Twenty hours each.

the *Escuela Nueva* (EN) model. It was created to sustain, expand, and improve on the EN model, as well as to continue innovating to offer quality education to the most disadvantaged children in Colombia and abroad.

In 1989, the World Bank selected the EN model as one of the three most effective education reforms influencing public policy design and implementation in developing countries (Mulkeen and Higgins 2009, Gaviria in press). From 1989 to 1995, the EN model was mainstreamed to public education and scaled up by the Ministry of Education, bringing it to 25,000 rural schools across Colombia as part of the country's Primary Education Universalization Plan. However, as a result of decentralization in the educational system in Colombia in the mid-1990s, the EN team at the Ministry of Education of Colombia was dismantled; the budget for EN's implementation was substantially decreased, and trained and experienced EN teachers were relocated and their working conditions worsened drastically (Colbert 2015, Gaviria, in press).

For Vicky Colbert, the decimation of the EN model in hands of the Ministry made evident the need to create an institution to sustain the model. In describing why she created the EN foundation, she explained, "I founded the organization because I saw that innovations are very vulnerable to political and administrative changes" (Hehenberger et al. 2010, 7). Thus, FEN was established as a nonprofit organization which, through its educational programs, would deliver services and advocate for quality of education for all.

FEN's work uses scaling-up as a strategy to effect change in education policy. It is a case of political advocacy (Jenkins 2006) because, through this strategy, EN influences education policy makers so that they will adopt its innovative education programs and take their implementation to scale. In fact, in 2002 at the World Education Forum in Dakar, it was acknowledged that *Escuela Nueva* was an example of how governments rely on third parties to guarantee basic education (UNESCO 2000). Moreover, FEN can also be considered a case of "NGOs as innovators" (Najam 1999), where NGOs create models to show governments different and new ways of achieving social impact. With *Círculos de Aprendizaje* as an example, it is clear that FEN developed an innovation that demonstrated to the government new ways to incorporate out-of-school children into the educational system.

FEN's initiatives to mainstream and expand its programs nationally and internationally are indicators of its advocacy strategy. As FEN implemented the EN model, it developed several innovations to increase the model's reach to urban areas (i.e., *Escuela Activa Urbana*), marginalized populations (i.e., *Círculos de Aprendizaje* and *Círculos para ENprender*), and other countries (i.e., international implementation strategy). Through its work, FEN aims to lead large-scale social change by mainstreaming its programs into existing governmental initiatives and by providing specialized expertise, technical assistance, and implementation support. FEN's impact is impressive due to its work with a wide range of partners, including governments, multilateral organizations, and private corporations. Indeed, in Colombia, there are other organizations that implement the model in a variety of regions; for example, the Coffee Grower Association of Caldas and the University of Pamplona, which promote the EN model in their respective areas of influence (Hehenberger et al. 2010, Colbert 2015). Internationally, FEN has facilitated the implementation of the model in Vietnam, Zambia, East Timor, Puebla State-Mexico, Peru, Dominican Republic, Brazil, Panama, Honduras, El Salvador, Nicaragua, and Guyana (Colbert 2015).

Círculos de Aprendizaje

Círculos de Aprendizaje originated in the midst of a dramatic increase in the number of internally displaced people (IDPs) in Colombia during the early 2000s due to the intensification of paramilitary groups and guerrilla actions (Acción Social 2010). In order help restore the right to education of internally displaced children, in 2001 FEN adapted the *Escuela Nueva* model with the aim of developing a flexible educational alternative for out-of-school children with an emphasis on IDPs (Fundación Escuela Nueva Volvamos a la Gente 2005). The U.S. Agency for International Development (USAID) and the International Organization for Migration (IOM) supported FEN's strategy through funding and technical support.

The pilot project of *Círculos de Aprendizaje* lasted 18 months, from May 2003 to November 2004, and was implemented in Altos de Cazucá, a neighborhood of Soacha in the Department of Cundinamarca. Findings demonstrate that the program effectively helped students to go back to the regular school due to the academic and socio-emotional support received in *Círculos de Aprendizaje* (Colbert 2015, Fundación Escuela Nueva Volvamos a la Gente 2005).

The Program's Approach

Inspired by the EN model, *Círculos de Aprendizaje* focuses on primary education in multigrade classrooms and with a learner-centered approach. It incorporates basic inputs of the EN model, where inputs are understood as the resources required in the education process to guarantee and improve knowledge and skills development (Behrman 2010, Barrow and Rouse 2005). Learning guides, adapted for *Círculos de Aprendizaje*; classroom instruments (e.g., self-reporting of attendance,[1] friendship mail,[2] traveling notebook,[3] and resources (e.g., learning corners,[4] a library); hexagonal desks; and teachers trained in the program are examples of educational inputs (see Colbert et al. 1993 and Schiefelbein 1993 for more detailed descriptions of the EN model).

To meet the needs of IDPs and other vulnerable children who have been out of the educational system for at least six months, *Círculos de Aprendizaje* adjusts to students' circumstances instead of demanding that students adjust to the program's tenets. Consequently, the program is located in marginalized neighborhoods where its students live, makes available a social worker or psychologist to help children and families with diverse issues, and provides children with all the necessary resources (i.e., school supplies, snack/lunch, transportation if needed). Although the program is located

---

1 In this tool, students mark their own attendance on a poster-size chart. This practice aims to strengthen students' punctuality, responsibility, and desire to go to school (Fundación Escuela Nueva Volvamos a la Gente 2016).
2 This is a communication tool in which students express their feelings and thoughts toward their classmates. It strengthens socio-affective ties among students, promoting coexistence, tolerance, and respect in the classroom; it also reinforces literacy skills (Ministerio de Educación Nacional 2006a, b, 2012).
3 This is a communication tool among students, tutors, and the community at large, especially parents or other people close to the program. The traveling notebook presents a topic and different actors have the opportunity to express their thoughts and feelings about it. It allows everyone involved in the process to learn about each other, strengthening respect, tolerance, and self-esteem (Fundación Escuela Nueva Volvamos a la Gente 2016, Ministerio de Educación Nacional 2012). The notebook goes from student to student making a journey across all students' homes or communities; the temporary keeper should make a contribution regarding the topic posed by the notebook.
4 Learning corners are spaces in the classroom that allow children to learn through experimentation as opposed to passively memorizing the contents of a book. Learning corners stimulate creativity, interaction between students, and offer opportunities for parents to work with their children(Fundación Escuela Nueva Volvamos a la Gente 2016).

outside a school, its students are formally enrolled in a traditional school, referred to as the *Institución Educativa Madre* (Mother School).

The *Círculos de Aprendizaje*'s general purpose is to reintegrate marginalized children into the educational system. As it is conceived as a temporary solution, the program operates on a year-to-year basis, which means that it changes its location every year and students stay in the program just for one academic year. When the academic year is finished, students are expected to continue their studies at the *Institución Educativa Madre*, which is a conventional school.[5] With the aim of reintegrating children into the educational system, the program is expected to ensure the development of literacy, mathematical logic, and socio-affective and citizenship skills to allow its students to perform and continue in traditional schools (Ministerio de Educación Nacional 2006b, 2012, Fundación Escuela Nueva Volvamos a la Gente 2009).

Taking the Program to Scale

In 2007, the Ministry of Education decided to mainstream and then implement *Círculos de Aprendizaje* at the national level of three factors. First, the Ministry had no educational alternative for addressing IDPs' educational needs. Second, *Círculos de Aprendizaje* demonstrated success during the pilot phase; and third, FEN's advocacy for the needs of internally displaced children was respected and widely acknowledged. As Vicky Colbert explained: "The Minister at the time had always been close to FEN, and had always believed in our work. . . . I took her to Altos de Cazucá to see the [*Círculos de Aprendizaje* classrooms], and she said: 'This is perfect; I want this for all the country.'"

The trajectory and legitimacy of FEN, including the familiarity of its executive director, Vicky Colbert, with the Ministry of Education, were critical factors in the scaling-up process. Thus, *Círculos de Aprendizaje* became an innovation that not only was incorporated by the Ministry as the strategy to address the needs of IDP children, but also influenced public policy by providing input to the policy guidelines for vulnerable populations' education (Gaviria in press).

The Program Implementation Process

Scaling up *Círculos de Aprendizaje* required a complex multilayered structure that involved different actors at different levels, yet FEN did not play a relevant role in this process. Due to the decentralized nature of the educational system in Colombia, the Ministry promoted and funded the program,

---

5 Parents could decide to enroll their children in a different school; in either case, students would continue in the educational system.

but it was implemented and administered by private nonprofit organizations. These organizations were selected through a public bidding system, based on their proposals for program implementation and demonstrated experience in the field of education.

Lack of understanding of the program by the organization selected to scale it up, a factor discussed in the next section, was a core reason why its implementation was hindered. The result was insufficient collaboration among actors and levels in the structure, misplaced priorities, and decontextualized actions.

At the national level, the main actors in the implementation process were the Ministry of Education, program operators, and the auditing firm charged with evaluating progress. At the regional level, actors included the regional teams, local Secretariats of Education, and the *Instituciones Educativas Madre* (IEM).

The Ministry of Education had a major role in ensuring the quality of the program and creating structures and processes so that *Círculos de Aprendizaje* could continue without the involvement of the national actors, such as program operators. The Ministry's responsibilities included funding the program, guaranteeing that it fulfilled its goals, and offering technical and administrative support. For example, the Ministry would offer appropriate assistance to program operators and regional Secretariats of Education when needed. In addition, in order to guarantee the quality of the education service, the Ministry had to monitor and retain comprehensive knowledge of the program based on the data collected, making the necessary adjustments if results did not meet established criteria (Ministerio de Educación Nacional 2006a).

Two of the program implementation actors—program operators at the national level and regional teams—were key to the process; they had major responsibilities, many of them shared with each other, for guaranteeing the sustainability of the program. Neither group was able to complete its work without the other.

Program operators were responsible for the delivery of *Círculos de Aprendizaje* into the classrooms (Ministerio de Educación Nacional 2011) and for providing all the necessary resources and guidelines for successful implementation at local level. They had to meet the Ministry's expectations and conditions. Their responsibilities ranged from funds management (i.e., payroll, additional contracts, acquisitions) to guaranteeing that students received, both academically and socially, what the program was meant to offer them.

Regional teams represented the program operators' facilitators in the regions. While program operators established guidelines and provided the necessary resources for implementation, the regional teams actually put *Círculos de Aprendizaje* into practice locally. Among many other activities, they coordinated daily classroom activities, built partnerships with local communities, and integrated the program to IEMs (Ministerio de Educación Nacional

2006a, 2011, 2012). Although program operators were responsible for the program as far as the Ministry was concerned, it was the regional teams that lead teachers training, provided continuous advice, and coordinated link-up among the program, IEMs, and Secretariats of Education in the regions.

Instituciones Educativas Madre had to guarantee that *Círculos de Aprendizaje*'s students were officially enrolled in the school during and after the program implementation, and ensure the continuity of the program within the school (Ministerio de Educación Nacional 2006a). Prioritizing education for marginalized and out-of-school children in general, and IDPs in particular, in schools' missions, was one of the crucial components of sustaining the program in schools.

It was also expected that, at the end of implementation, IEMs would have the capacity to run *Círculos de Aprendizaje* classrooms without the external support of the regional actors. In fact, all of the program inputs, except the teachers, were transferred to the IEM at the end of each academic year.

Local Secretariats of Education had to support IEMs' efforts to integrate the program into the school and offer assistance to program operators and regional teams as needed (Ministerio de Educación Nacional 2006a, 2011). Further, they had to build necessary institutional alliances to expand the program to other local schools in addition to the original IEM. IEMs and local Secretariats of Education were responsible for integrating the program and its principles into their work. The depth of the implementation was strongly related to the work of these two actors.

Finally, the auditing firm had to verify the satisfaction of every obligation stipulated in the contract between the Ministry of Education and program operators. It was the auditors' responsibility to approve the Ministry's payments to the operators and identify strengths and weaknesses in the program's implementation (Ministerio de Educación Nacional 2011). As its representatives visited the classrooms, it became an important actor during the implementation process. The Figure 4.1 summarizes the *Círculos de Aprendizaje* implementation structure.

As shown in Figure 4.1, interactions between actors were essentially hierarchical, with the actors on a national level directing the work in the regions. As explained in the following section, the nature of these relationships was a major challenge faced in scaling the program at the national level.

The Role of Fundación Escuela Nueva *in Implementation*

As noted above, when understanding *Círculos de Aprendizaje*'s implementation at scale, it is important to be aware that FEN was not part of that process. Although during the design stage FEN did play a role by providing technical assistance on the objectives, activities, and expected outcomes of the program, once mainstreamed, the Ministry did not incorporate FEN as a provider of specialized expertise, nor did *Escuela Nueva* assume such a role. During the first year that the Ministry implemented the program at scale,

Círculos de Aprendizaje 93

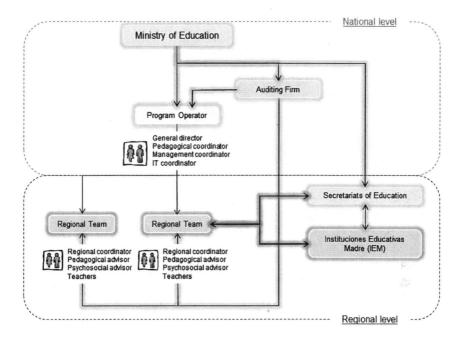

Figure 4.1 Círculos de Aprendizaje's implementation structure

FEN offered training and technical advice to program operators, but it did not continue in the subsequent five years.

Following similar procedures used to implement its programs internationally, FEN, in preparation for the scaling-up process in Colombia, handed over all the necessary documentation regarding *Círculos de Aprendizaje* to the Ministry, helping it to delineate the scaling-up strategy. Nevertheless, once the program was mainstreamed and scaled up, FEN did not request a permanent role in the program. In fact, after the implementation went to scale, the only possibility that the Ministry provided to FEN was to become a program operator, meaning FEN would have had to participate in the public bidding process, along with other education organizations with no previous expertise with the program.

The Challenges of Scaling-Up Círculos de Aprendizaje

Implementation at scale of *Círculos de Aprendizaje* evidenced challenges at different levels that prevented the innovation from transforming the education system and that raised questions about the appropriateness of the structure designed for its implementation and negatively affected the program's quality and sustainability.

This section explores two major challenges faced by education innovation when scaling-up: ownership and depth. The transfer of project ownership refers to the shift when an innovation moves from an external/imposed reform to an internal/locally ingrained process (Coburn 2003). It involves a change of authority, and the new owner's development of the necessary capacities and skills required to maintain, improve, and expand the innovation, as well as collaboration between all actors involved in the implementation (Coburn 2003, Elias et al. 2003). Thus, the shift of ownership directly affects the sustainability of the program and ultimately its quality. In the case of *Círculos de Aprendizaje*, ownership would imply that regional Secretariats of Education and IEM schools could implement the program by themselves with the right support from the Ministry of Education, but without the authoritative intervention of third parties such as the program operators or regional teams.

Depth implies that the implementation of the innovation "goes beyond surface structures or procedures [and alters] teachers' beliefs, norms of social interaction, and pedagogical principles as enacted in the curriculum" (Coburn 2003, 4). Thus, depth, in scaling-up *Círculos de Aprendizaje*, requires that implementation goes beyond input provision (i.e., learning guides, desks, supplies) to the creation of a new mindset in teachers, schools, and regional Secretariats of Education that encourages the successful integration of out-of-school students. It requires incorporating such purpose and mindset in schools mission statements, thus resulting in improved teaching and learning in classrooms and schools. As with ownership, depth affects the sustainability of the innovation, but equally important, it affects its quality (Coburn 2003).

Challenges about ownership and depth are linked, first, to the multilayered and complex structure designed for the implementation of the program at scale. Second, challenges affect the vision about what success would look like for students, classrooms, and schools, and the conditions to achieve such success. As a result, everyday difficulties in the classroom, from the use of the learning guides to the teaching strategies utilized, were highly influenced by the characteristics of the implementation process.

During its scaling-up at national level, *Círculos de Aprendizaje* faced a disconnection between the way the program was originally designed by FEN, the modification implemented by the Ministry, and the reality of the regions. This resulted in misplaced priorities and miscommunications between actors.

Discord Among the Ministry, Program Operators, and Regional Teams

Interviews and observations revealed that members of the regional teams did not function as a unified body; instead, they struggled individually with their responsibilities, in an environment of mutual distrust and/or alienating demonstrations of power.

Regional teams often implied they did not have a solid working team through, for example, remarks about not receiving support or advice from program operators, not having reliable coworkers, and autocratic behavior by national and regional coordinators. Overall, participants questioned the commitment of other colleagues toward the program, the children, and the team.

An authority unbalance between the national (i.e., the Ministry, program operators, auditors) and regional level (i.e., regional teams, IEMs, Secretariats of Education), and among members of regional teams was evident. From the top down, the Ministry and auditors had specific requirements for program operators; the program operators, in turn, translated those requirements to specific demands for regional teams; regional coordinators passed on those demands to pedagogical and psychosocial advisors who then made them obligations for the teachers. Regional team members regularly expressed confusion over, or disagreement with, commands from a higher authority. For example, "I don't know why but Bogotá [program operators, the Ministry, or auditors] wants these forms [filled out] every week. Now we have to sing and play all the time because the general director said so" (Alexa, regional team coordinator). Or, "kids can't take [snack leftovers] home because the coordinator doesn't like it" (Daniel, program's teacher).

Collaboration is a key factor when scaling-up an innovation (Fishman 2005), and it becomes essential during the buy-in process when the innovation reaches local areas beyond the central and regional level. A collaborative process helps adopters to understand the innovation and developers to understand the reality of the context in which the innovation is implemented. Instead, with *Círculos de Aprendizaje*, as communication between levels and among team members was sporadic and unfriendly, the comprehension of the program was limited and the possibility of developing ownership at the local level was limited. Threats, accusations, and orders were rather frequent among actors. Therefore, organizational relationships were strongly based on control and supervision instead of on support and guidance. Teachers were especially affected by these relationships as they were responsible for what happened in the classroom.

Providing continuous support and advice was one of the most important responsibilities of the Ministry, program operators, and regional teams. The Ministry needed to facilitate program operators' work and provide training and technical advice about the program. Program operators were responsible for training and ongoing pedagogical qualification of regional teams. Regional teams had specific members in charge of supporting teachers daily, which implied training and continuous advice about the pedagogical implications of the program in the classroom. Nevertheless, advisory roles changed from one that offers continuous support and advice to one that controls and supervises.

Inadequacies of the Ministry of Education

The Ministry's shortcomings not only affected the possibility for other actors to understand, and in consequence, to buy-in to the program; they also affected the possibility of building the required capacities to sustain *Círculos de Aprendizaje* in the long term.

The Ministry monitored program implementation through an audit firm. The auditing process was described as meticulous and was considered by the Ministry "the main source of information about the implementation of *Círculos de Aprendizaje*" (Diego, representative of the Ministry). Still, the Ministry seemed unaware of critical regional issues.

Teachers, coordinators, and Secretariats of Education representatives, who had been involved in the program for several years, asserted that the literacy struggles of *Círculos de Aprendizaje* students had been present since the program was first implemented across the country. Moreover, auditors not only acknowledged this situation but stated that it had been systematically described in their reports year after year.[6] However, the Ministry had no strategy in place to address this issue, either improving the program materials, building the capacities of Secretariats of Education and schools to address it, or giving special guidelines and training to program operators.

In addition, the Ministry seemed overly confident about program operators' capacities and achievements. Juliana, a representative of the Ministry, explained: "The audit firm reports to us if contractual obligations are being fulfilled, we have the reports, lists, and so forth . . . so we know everything is in order." This situation was aggravated by the fact that the Ministry did not have a follow-up system to assess if children continued at school, the main indicator of program success. When asked if the Ministry knew the success rate of the program, in terms of children reintegrated into the educational system, a Ministry representative replied that they did not have that information.

Limitations of the Program Operators

The work of program operators in the regions exhibited the same limitations the Ministry harbored at the national level. Most procedures designed to effect the implementation and guarantee its sustainability were inadequate. Moreover, they made evident a lack of coherence between requirements and goals made at different levels regarding the program's implementation. Program operators were not able to contribute much to the shift of authority in the regions, nor to the depth of the implementation, thereby directly compromising the sustainability and quality of the program.

Observations and the analysis of program reports showed that the operators played a fundamental role in terms of logistics and supervision, but

---

6 The auditing process was conducted by the same group each year.

linked to the findings reported above, the regional teams, Secretariats of Education, and IEMs did not receive enough resources or assistance to fulfill their role effectively. In fact, in different regions, Secretariats of Education members repeatedly said that program operators "were caught up in technical and logistical issues, far from the human dimensions of the project and overlooking the real needs of the children."

When asked about strategies to support the multiple regions simultaneously and improve local pedagogical skills, Juan, a program operator team member, explained:

> We started a process, you tell people [regional teams members]: "this is what we have to do," but it doesn't work, it doesn't work anywhere in the country. If you create a school right here, well, that's a different story, but working long distance, there's nothing you can do. . . . I've traveled maybe a couple of times to some regions, but not more, there are regions I don't even know yet. I mean, if the general director needs me to go somewhere, I'll go, but it's not like I go to the regions regularly. . . . When I've been there, I've spoken with the pedagogical advisors, and you expect that they will implement what you just told them, but who knows, maybe they don't even understand what you are telling them.

In addition to program operators' difficulties in effectively supporting the implementation at the regional level, program operators and regional teams' members also reported that the organizational structure of program operators did not empower staff autonomy. Juan explained, "We don't decide anything here, we only follow orders from the general director."

Finally, work conditions at the level of the program operators did not always foster the professionalism required to support the regional teams. For example, regional team members mentioned a "total lack of politeness" (Ángel, regional team member) and injustices in the way they were treated by the program operator:

> How can you talk about equity, about rights, when you have a lousy salary? Or when they [program operators] pay you two weeks late? Or tell you there's no money to pay you? Or they retain your salary because one of the forms has an error? I mean, with their actions, [program operators] are going against the ideas they are selling with *Círculos de Aprendizaje*.
>
> (Alexa, regional team coordinator)

Lack of Ownership by the Regional Teams

It can be argued that regional teams were the most important actors when implementing *Círculos de Aprendizaje*. They were responsible for everything that happened in the *Círculos de Aprendizaje* classrooms, for the program's success in terms of children's achievements, and for sustainability of the implementation. Still, data show they did not have much, if any, ownership of the program or competencies to implement the program in depth (i.e., beyond educational inputs).

Organizational relationships observed for program operators were replicated within the regional teams, although the extent of teamwork varied depending on the way coordinators exercised their role in terms of their management style and involvement with students and classrooms. Rather than generating a shared culture and standard approach of the program across regions, management styles ranged from autocratic, where the exercise of power, control, and supervision of staff members were the norm, to permissive, where coordinators did not regulate or intervene much in the staff's performance. Involvement in the classroom varied between high involvement, meaning that regional teams' coordinators were engaged in daily activities, challenges, and needs of students and classrooms, to low involvement, meaning that coordinators were not involved with issues concerning the daily lives of students or the classrooms.

The amount of effective work time in the classroom, teachers' punctuality, general routines in the classrooms, and even students' attendance varied depending on the type of coordinator. The region that worked better overall had a hands-on coordinator, highly involved in the classroom daily activities, as opposed to a desk coordinator. Nevertheless, understanding the program and its pedagogical implications was limited. Camila, a member of a regional team, expressed:

> The program's methodology is not so clear; the truth is I don't totally understand what the program is about. The fact is that the program's operator is not very clear on certain things and focuses more on classroom instruments, and I haven't tried to understand it on my own either.

It was reported that pedagogical advisors, the individuals in charge of offering continuous pedagogical support to teachers, did not offer proper help. Doing so would have positively affected the depth of the implementation at scale, defined above as the extent to which an external reform causes positive change in classroom practices. On the contrary, there were multiple conflicts between teachers and advisors. Teachers stated: "We didn't have proper training . . . [and] when we have questions or say we don't understand something, his [pedagogical advisor] answer is 'you're supposed to be professionals, aren't you teachers?, so, you figure it out'" (Daniel, program teacher). Valeria explained, "The pedagogical advisor doesn't go to the classrooms, we pretty much never see him so we do what we can on our own," and Benjamín, a progam teacher, added, "Our meetings [are used] to lecture us [teachers], we don't have much training or advice, the pedagogical advisor goes with her checklist to the classrooms and that's it. We have to get creative to solve our problems."

Observations supported teachers' views, and it became evident that teachers did not receive support on a regular basis or have proper basic training at the beginning of the program. Even teachers with previous experience in *Círculos de Aprendizaje* exhibited difficulties when facing challenges in the classroom. In consequence, teachers decided by themselves the criteria for evaluating the students, how to organize the classwork, and, in some cases, how to use the program's tools, resulting in lack of consistency across classrooms.

Misplaced Priorities

Priorities reflected an emphasis on breadth instead on depth during the scaling-up process of *Círculos de Aprendizaje*. It has been extensively discussed in the literature that most scaling-up efforts focus on breadth, which refers to the quantitative expansion of the innovation rather than the intensity and effectiveness of reform efforts at local level (Robinson et al. 2016). Although breadth and quantitative expansion are essential parts of the diffusion of innovations across contexts, it is also paramount to focus on the quality and depth of implementation. For example, interviews showed that program priorities—such as quantitative coverage, logistical program tasks, educational inputs such as learning guides for students, and budgets—were not entirely aligned with the program's purpose, which was the reintegration of out-of-school children into schools and to modify pedagogical practices in classrooms to support such reintegration. This misalignment in turn made the buy-in process by local implementers even more difficult. In the case of *Círculos de Aprendizaje*, the feeling of unfairness certainly did not help teachers, and other team members, to understand the program and become its advocates.

Interviewees continuously stated that program implementation prioritized quantitative aspects such as the number of participating students attending school, as well as other concrete, but partial, numerical indicators over students' learning processes, pedagogical needs, and regional particularities. For instance, Victoria, an IEM teacher, said, "The Ministry demands full coverage, but here it is just too complicated, kids come and go all the time and there's nothing you can do about it," and Andrea, a program teacher, explained, "When we started a few years ago [implementing the program in the region] . . . we had smaller groups—15, 16 kids—now sometimes we have up to 25! How can you give them all what they need, which, by the way, is a lot?"

Emphasis on coverage virtually determined the program's quality of implementation in the regions. Most of the energy and time was invested in keeping the classrooms full: Teachers were constantly trying to explain why there was a variation in students' attendance; or making phone calls, summoning parents, visiting homes, and/or looking for new students. Coverage was important in terms of increasing the extent of program's results (i.e., more children reintegrated into the school system). However, it was done at the expense of overseeing the quality of teaching practices in classrooms. Such attention to quality would have impacted positively in the depth dimension of the program, including practices in classrooms for students attending school.

The lack of attention to program quality and the educational process within classrooms described above resulted in a disengagement from students' learning processes, exacerbated by the importance placed on logistical and administrative tasks, as expressed by interviewees: "what's really important here is that I have to hand over eight meetings' minutes, a bunch of photographs, and the tons of forms they have [program operators and the Ministry] filled out to perfection, that's it, that's what's important in *Círculos de Aprendizaje*" (Ángel, regional team member). Regional team members, mainly teachers, had to fill out, among other forms (1) daily reports of attendance; (2) daily reports of meals distributed with children's signatures or fingerprints; and (3) weekly reports of dropouts or missing students, including the follow-up of each case. These forms were, in fact, a contractual obligation, but in the same way that coverage does not incorporate the idea of process. Sebastián, a representative of a program operator, stated:

> [The forms] don't reflect either the process that is supposedly expected from the program nor any concern, or knowledge, about the pedagogical core of *Círculos de Aprendizaje* . . . contracts were probably made by experts in contractual processes but they clearly know nothing about pedagogy or perhaps education, so, they . . . leave out indicators of any process with the students or the community.

Educational inputs were used as success indicators for program monitoring. In the case of *Círculos de Aprendizaje,* these inputs were teachers, learning guides, classroom instruments and resources, among others. This implementation model is driven exclusively by education inputs such as distribution of learning guides to schools as opposed to educational processes and interactive practices in classrooms. The model allowed the Ministry and program operators to easily report that the program was being implemented correctly, leaving attention to quality aside. For example, the focus on inputs ignored attention to the interactive practices between teachers and students that support quality of learning in classrooms. According to María, a program teacher, visits by the auditors and even other members of the regional team were "a verification of a checklist, they will get very mad if you don't have all the classroom instruments on the walls, or if you're not using the learning guides."

Furthermore, this emphasis on inputs and a lack of a culture of continuous improvement promoted questionable behaviors focused on satisfying the auditors. For example, teachers altered classroom instruments or other materials important to auditors so that auditors would evaluate them positively when visiting their classrooms. Regional and national staff members hid or overlooked major difficulties within the teams or adjusted reports and forms. It was also reported by a staff member that, in one of the initial training workshops, the program operator invested a day on teaching regional teams how to correctly answer auditors' questions.

Finally, participants said that program operators' profit-making interests determined the program's overall implementation at scale rather than an interest in students or regional team members' needs, as seen in the physical conditions of classrooms and the working conditions of regional teams' members. Aside from children's satisfaction with their classrooms, the actual states of most of these spaces were unsatisfactory, even "anti-pedagogical" as an IEM teacher called them. They were inadequate in size, light, ventilation, and security; and three of four regions did not have drinkable water. Problems with the classrooms were related to the fact that there were few options to choose from, considering the neighborhoods where *Círculos de Aprendizaje* was implemented. However, some regional teams' members stated that program operators did not approve a larger budget in order to rent a better house or make the necessary adjustments to improve the house that was rented. This lack of attention by program operators and the Ministry in ensuring a minimum of basic school infrastructure—indeed, one of the main resources required to achieve success in the educational process—contrasts with the excessive attention to the *Círculos de Aprendizaje* program inputs.

People in the regions, particularly teachers, had long workdays and low wages. Even though they were thankful for having a job and the opportunity to work with vulnerable children, most participants stated their working conditions were unsatisfactory. The majority of teachers worked two school periods per day, each five hours long. The teachers had free time of 30 minutes or an hour maximum between the two periods. Their monthly pay was about US$ 600.[7] Wages did not differ much between operators, although contract conditions were completely different; one program operator hired teachers as consultants, the other one as employees.[8] In general, teachers, coordinators, advisors, and representatives of Secretariats of Education agreed that working two school periods per day was excessive, but the decision to have teachers working all day in the classroom was made by program operators. Even though program operators asserted that double shifts existed because of technical requirements, this was not a contractual

---

7 COLP $1,200,000.00; This was the average of wages offered by program operators, including the social security fee that every person must pay by law. In 2012 the exchange rate was US$ 1,00 = COLP $ 2,000.00.
8 These contractual differences, consultant vs employee, represented important differences in terms of the social security benefits, as well as the stability offered to the teachers.

obligation or an instruction from the Ministry. In reality, double shifts seemed to represent a financial gain for program operators. Maya, a representative of a Secretariat of Education explained:

> What is the rationale behind the double shifts? Well, first, a teacher's wage is really bad, it's dreadful, really dreadful . . . so program operators disguise how bad the pay is by packing together two school periods. It's a strategy to entice teachers; the poor teachers feel they have a semi-decent salary but they have to work all day long for it, all day, with two different groups, being tired, so productivity in the classroom is low and these kids' progress, well, is not good either. Besides, they have tons of other responsibilities so teachers often limit their teaching time, and that's time that children are not working. Poor salaries, long days of work, and why? Because program operators save money, they don't have to pay extra time, for example. So, what's the operator looking for? Not students' or teachers' wellbeing, but to make more money.

Disconnection From Local Realities

There was a general consensus that the program was needed in the regions, but interviewees expressed that the regions did not have a say in the program implementation, which affected local ownership and the sustainability of the program. One of the disconnects supported by the evidence is that the scaling-up process did not correspond to the reality of the contexts in which the innovation was to be implemented. For instance, in one region, there were not enough out of school children for the program. Additionally, the structures and supports put in place were inadequate and limited. For example, Secretariats of Education did not report having requested participation in the implementation of the *Círculos de Aprendizaje* program, nor did they agree with the way it was rolled out. Thus, the possibilities for the innovation to impact the regional policy or the institutional structure of the IEMs were minimal.

The Ministry made decisions and demands regarding the program without full knowledge of regional particularities, which affected the Ministry's legitimacy as a guarantor of the program's quality. For example, José, a member of a Secretariat of Education, said: "We have kilometers and kilometers of water, and they [the Ministry] always ask if there's any elevated ground to place the classrooms! . . . They really don't know what they're talking about!"

A second disconnect was that the Ministry was unable to account for the quality of the program at the regional level. Juliana, a Ministry representative, stated:

> I have the reports, pictures, and records that state that teachers and IEM teachers were trained in *Círculos de Aprendizaje* . . . [but] I can't

tell anything about these people's knowledge after the training, I could not say if they are learning or not, I don't know if we evaluate that.

If output evidence (e.g., attendance lists, photographs, or meeting/workshops reports) was available, it was assumed that everything in the program was working as expected. However, when asked if the activities were achieving what they were intended to do, some Ministry's representatives said they could not tell anything beyond the evidence while others openly said, "you have the documents and everything, but you know the program is not working."

The Ministry's process for improving *Círculos de Aprendizaje*, which was inadequate and came late in the scaling-up of the program, is the third element that illustrates the disconnection between the Ministry and program implementation. First, it took more than five years of implementation before the Ministry issued an opinion of the program (Ministerio de Educación Nacional 2013), which was not a formal evaluation but an assessment of its coherence and relevance according to the country's regulations and standards. Second, the Ministry's opinion was issued exclusively on the basis of materials and documents; lacked in feedback from the regions. People involved in program evaluation were not aware that children in the regions had severe literacy problems, and thus they were not able to work independently with the learning guides. Thus, although the Ministry's opinion highlighted conceptual weaknesses in the learning guides, in general, it stated that the program was adequate and was producing the expected results.[9]

Finally, observations and interviews evidenced a lack of understanding between program implementation and reality in terms of program operators' expertise. Although the selection process was characterized as strict, it seemed that program operators did not have what was required to guarantee adequate implementation of *Círculos de Aprendizaje*. As was said by different participants, *el papel lo aguanta todo* (anything and everything can be said on paper), meaning operators' staff members did not meet expectations. Juan, a representative of a program operator, explained:

> You know, everything also depends on our professional experience. I mean, to tell you the truth, I, for example, knew something about these kind of programs [flexible educational models], and I had worked a little on the pedagogical part, but it was more theoretical, and that's very different from being in charge of actual processes in other regions, to be responsible and aware of what's happening and what's not.

---

9 Assessment of results are also based on what is stated in official documents such as the pilot project report (Fundación Escuela Nueva Volvamos a la Gente 2005) or the implementation's manual (Ministerio de Educación Nacional 2006a).

Lessons Learned by *Fundación Escuela Nueva Volvamos a la Gente*

Since the mid-1990s, FEN and its CEO Vicky Colbert knew about the fragility of innovations under government leadership. The process of scaling-up *Círculos de Aprendizaje* became one more example of the difficulties of expanding an NGO's impact by working together with governments.

For FEN, it is clear that to impact educational change, it is necessary to influence public policy, and doing so only plausible if the challenges of working with Ministries of Education can be surmounted. Aiming to overcome the obstacles faced by the organization in the past when scaling-up its innovations, FEN restructured some of its strategies. For instance, FEN and its allies now create memorandums of understanding that include the scope of the collaboration, the conditions needed to maintain the quality of the innovation, and the responsibilities of each partner. In addition, there is a copyright agreement that seeks not only to protect FEN's authorship, but also the quality of the materials and processes involved in the innovation. Finally, to help to oversee the quality and sustainability of the innovation in the long term, FEN seeks to include in the collaboration another organization, whether a strong, well-consolidated NGO or a company in the region.

Yet, these changes need to be tested in a process of the scale of the national implementation of *Círculos de Aprendizaje*. Unfortunately, it is possible that they might not work, particularly when partnering with foreign countries, because follow-up and accountability processes are difficult to reinforce.

Conclusion

The case of the *Círculos de Aprendizaje* illustrates the major difficulties an innovation faces when taken to scale, particularly when the scaling-up process involves different actors at different levels. As recounted in this chapter, the case shows in detail the main obstacles to the creation of the necessary conditions for delivering a quality program and guaranteeing its sustainability.

As Coburn (2003) discusses, a scaling-up process requires a combination of depth and breadth that allows the entire system to change. With respect to education, it is necessary that teachers, schools, districts, and even the community change in order to sustain the innovation and the gains it offers in a certain context and for a specific population. A key element when scaling up is building capacity in all people and institutions involved in the innovation.

As evidence from the *Círculos de Aprendizaje* scaling-up process shows, most components of the education system were not transformed by the innovation. On the contrary, comprehension of the program itself and the capacities needed to correctly implement and sustain it were low. There was clearly a disconnection between the levels of involvement in terms of the actors and their priorities, and the way that the entire scaling-up process functioned in different contexts. It had been necessary to transcend the focus on educational inputs to invest time and resources to continuous training and support processes to improve outcomes, but these investments were not made.

Consequently, the role played by the Ministry of Education in Colombia needs to be reexamined in terms of its responsibility to ensure the program's quality and capacity building at different levels of the system. Further research can analyze this issue in the context of a decentralized system such as that in Colombia.

Finally, it is necessary to consider the role that *Fundación Escuela Nueva Volvamos a la Gente* (FEN) should have in the scaling-up process of its innovations in the future. As Edwards and Hulme (1992) state, NGOs must restructure their organizational strategies when collaborating with the government. In the case of an organization such as FEN, it would be worth studying the conditions it could demand when one of its programs is taken to scale by a third party. The scaling-up experience of *Círculos de Aprendizaje* shows that FEN would have been a qualified advisor not only for regional teams, but for the Ministry itself, and with its assistance and support, the scaling-up process would have had better results.

References

Acción Social. 2010. *Desplazamiento forzado en Colombia*. Bogotá: Agencia Presidencial para la Acción Social y la Cooperación Internacional de la Presidencia de la República de Colombia.

Barrow, Lisa and Cecilia Elena Rouse. 2005. *Causality, Causality, Causality: The View of Education Inputs and Outputs from Economics*. Chicago, IL: Federal Reserve Bank of Chicago.

Bartlett, Lesley and Frances Vavrus. 2009. "Introduction: Knowing, comparatively." In *Critical Approaches to Comparative Education*. edited by Frances Vavrus and Lesley Bartlett, 1–18. New York: Palgrave MacMillan.

Behrman, Jere. 2010. "Investment in Education Inputs and Incentives." In *Handbook of Development Economics*. edited by Dani Rodrik and Mark Rosenzweig, 4883–4975 Elsevier.

Coburn, Cynthia E. 2003. "Rethinking Scale: Moving Beyond Numbers to Deep and Lasting Change." *Educational Researcher* 32(6): 3–12.

Colbert, Vicky. 2015. "Escuela Nueva: Una contribución a la calidad y a la equidad en la educación para el Siglo XXI." In *Equidad; Perspectivas para Colombia*. edited by Óscar Echeverri Cardona. Santiago de Cali: Fundación para la Educación y Desarrollo Social (FES).

Colbert, Vicky, Clemencia Chiappe and Jairo Arboleda. 1993. "The New School program: More and better primary education for children in rural areas in Colombia." In *Effective Schools in Developing Countries*. edited by Henry M. Levin and Marlaine E. Lockhedd. London, UK: The Falmer Press.

Edwards, Michael and David Hulme. 1992. "Scaling up NGO Impact on Development: Learning from Experience." *Development in Practice* 2(2):77–91.
Elias, Maurice J., Joseph E. Zins, Patricia A. Graczyk and Roger P. Weissberg. 2003. "Implementation, Sustainability, and Scaling up of Social-Emotional and Academic Innovations in Public Schools." *School Psychology Review* 32(3): 303–319.
Fishman, Barry, J. 2005. "Adapting innovations to particular contexts of use: A collaborative framework." In *Scaling up Success: Lessons Learned from Technology-Based Educational Improvement.* edited by Chris Dede, James P. Honan and Laurence C. Peters. Indianapolis, IN: Jossey-Bass.
Fundación Escuela Nueva Volvamos a la Gente. 2005. *Modelo articulado Escuela Nueva—Círculos de aprendizaje: Sistematización de la experiencia Fase I. Bogotá.* Colombia: Fundación Escuela Nueva Volvamos a la Gente.
Fundación Escuela Nueva Volvamos a la Gente. 2009. *Education in Emergencies: The success story of Learning Circles in Colombia.* Bogotá: Fundación Escuela Nueva Volvamos a la Gente.
Fundación Escuela Nueva Volvamos a la Gente. 2016. *Escuela Nueva Activa: Manual para el docente.* Bogotá: Fundación Escuela Nueva Volvamos a la Gente.
Gaviria, María Cristina. in press. *Historia de Escuela Nueva en Colombia: Una propuesta por las escuelas, política nacional de renovación pedagógica para la educación de calidad.* Bogotá, Colombia: Fundación Escuela Nueva Volvamos a la Gente.
Hehenberger, Lisa, Nathalie Moral and Johanna Mair. 2010. *International Expansion of Escuela Nueva: A Transformative Pedagogy on a Global Scale.* IESE Publishing.
Jenkins, Craig. 2006. "Nonprofit Organizations and Political Advocacy." In *The Nonprofit Sector: A Research Handbook.* ed. Richard Steinberg and Walter Powell. New Haven: Yale University Press.
Maxwell, Joseph A. 2005. *Qualitative Research Design: An Interactive Approach.* Thousand Oaks: Sage Publications.
Ministerio de Educación Nacional. 2006a. *Documento de apoyo a la replicabilidad del modelo Círculos de Aprendizaje—Escuela Nueva Activa (CA-ENA).* Bogotá: Ministerio de Educación Nacional.
Ministerio de Educación Nacional. 2006b. *Manual para tutores y tutoras del modelo de atención círculos de aprendizaje.* Restricted circulation.
Ministerio de Educación Nacional. 2011. *Proyecto de pliego de condiciones: Licitación pública LP-MEN- 11 DE 2011.* Bogotá, Colombia: Ministerio de Educación Nacional.
Ministerio de Educación Nacional. 2012. *Anexo técnico, administrativo, pedagógico y operativo del proceso de implementación de los modelos educativos flexibles del Ministerio de Educación Nacional.* Bogotá: Ministerio de Educación Nacional.
Ministerio de Educación Nacional. 2013. *Concepto de calidad modelo educativo flexible: Círculos de Aprendizaje.* Bogotá, Colombia: Ministerio de Educación Nacional.
Mulkeen, Aidan G., and Cathal Higgins. 2009. *Multigrade teaching in Sub-Saharan Africa: Lessons from Uganda, Senegal, and the Gambia,* World Bank Working Paper No. 173. African Human Development Series. Washington,D.C.: World Bank.
Najam, Adil. 1999. "Citizen Organizations as Policy Entrepreneurs." In *International Perspectives on Voluntary Action: Reshaping the Third Sector.* ed. David Lewis. London: Earthscan.

Robinson, Jenny Perlman, Rebecca Winthrop and Eileen Mcgivney. 2016. *Millions Learning: Scaling up Quality Education in Developing Countries* Washington, DC: The Brookings Institution.

Schiefelbein, Ernesto. 1993. *En busca de la escuela del siglo XXI: ¿Puede darnos la pista la Escuela Nueva de Colombia?* Chile: UNESCO/UNICEF.

UNESCO. 2000. World Education Forum. Dakar, Sengal, 2000. *Final Report.* Paris, France: UNESCO.

Vega-Chaparro, Laura María. 2014. "Círculos de Aprendizaje: Challenges and possibilities of flexible educational models for marginalized populations in Colombia." Doctor of Education Dissertation, International and Transcultural Studies, Teachers College, Columbia University.

Wils, Frits. 1995. "Scaling-up, mainstreaming and accountability: the challenge for NGOs." In *Non-governmental Organisations: Performance and Accountability Beyond the Magic Bullet.* edited by Michael Edwards and David Hulme, 53–62. London, UK: Earthscan.

5 Social Advocacy in Neoliberal Times
Non-governmental Organizations in Ecuador's Refugee Landscape

Diana Rodríguez-Gómez

Advocacy has been traditionally portrayed in a heroic light, rather than related to the social and political forces of neoliberalism. This chapter illuminates the challenges and tensions of social advocacy in a unique scenario where Ecuadorian non-governmental organizations (NGOs) navigate the contradictions between President Rafael Correa Delgado's nationalist, antineoliberal agenda and the demands of international development where many aspects of the dominant neoliberal agenda prevail. I define social advocacy here as the series of efforts that facilitate the translation of key messages across actors with unequal forms of power.

While some advocacy efforts take place at the global scale, as we have seen in the recent Syrian refugee crisis, NGOs in Ecuador are engaging in local efforts to assist adults and children in need of international protection. This study addresses local advocacy efforts in Quito, Ecuador, that seek to empower and raise awareness on the situation of marginalized youth and children, more specifically those who face the consequences of forced migration.

This country of 15.74 million inhabitants receives the largest share of refugees in Latin America and the Caribbean (United Nations High Commissioner for Refugees, [UNHCR] 2017a). In January 2017, the UNHCR reported a total of 60,520 individuals with refugee status in Ecuador, 95 percent of whom are Colombian (UNHCR 2017a). Nearly 60 years of internal armed conflict among left-wing guerrillas, right-wing paramilitary forces, and the Colombian national army have resulted in the forced migration of approximately seven million Colombians, both within and across national borders.[1] UNHCR (2017a) has also calculated that even after the Peace Accord between the Colombian government and the Revolutionary Armed

---

1 A 2011 report surveyed 1,500 individuals with refugee status in Quito and Guayaquil, and found that approximately 60 percent of the population with refugee status lived in urban areas, with 30 percent living in Quito alone. Not surprisingly, survey respondents reported the presence of armed actors, threats, personal attacks, and fear as causes for migration. The reasons cited for choosing Ecuador over other neighboring countries were geographical closeness, the presence of supporting social networks of friends, and the general perception of Ecuador as a peaceful country (Ospina and Santacruz 2011).

Forces of Colombia (FARC) was passed in November 2016, about 420 Colombians still cross the Colombia–Ecuador border every month in search of international protection. In 2000, to serve the needs of these migrants, Ecuador's government requested UNHCR's presence. Since then, to support UNHCR's work in the field, the most widely recognized refugee NGOs have been working in Ecuador.

This chapter presents an analysis of how conditions created by the Correa Delgado government and the prevalence of a neoliberal logic has shaped NGOs' advocacy efforts in Quito, Ecuador's capital. For the purpose of the discussion, I define neoliberalism as a set of ideas supported by practices that disseminate and naturalize economic rationality—the logic that prioritizes the efficiency of the free market—across both the private and public realms of life (Ferguson 2009).

The findings from this chapter are based on ethnographic research I conducted from May 2013 to June 2014 on NGOs in Quito. My investigation includes participant observation in two international NGOs, and participation in one advocacy undertaking: an art exhibition, named Flows, organized by a local NGO that sought to raise awareness about the life trajectories of Colombian youth with refugee status. This case exemplifies neoliberalism's potential to shape social advocacy, even when Correa Delgado's government claimed to implement the socialism of the twenty-first century in Ecuador.

In the next section, I discuss how the policies and activities of NGOs are impacted by the need to operate within a neoliberal context. Following that review, I draw from the interviews with NGO participants conducted for my study to present a picture of the refugee-aid sphere in Ecuador during Correa Delgado's government, and then offer my interpretation of social advocacy in a neoliberal context. After describing the methods employed to collect and analyze data for this study, I present the study's findings and conclusions, suggesting that the pressure exerted over the non-governmental sector during Correa Delgado's presidency created the conditions for NGOs to implement soft forms of social advocacy.

A Review of Research on Neoliberalism and its Effect on Social Advocacy by NGOs

Discussions about neoliberalism and the role of NGOs are not new to the field of comparative and international education (CIE). Within this field of inquiry, three strands of work can be identified. One tackles the effects of the provision of education by non-state actors and the possibilities of improving physical access to school for the most marginalized children (Blum 2009; Lewin and Sayed 2005; Nishimuko 2009; Rose 2007, 2009a, 2009b). A second line of research explores the relationships that emerge between NGOs and the state under neoliberalism (Archer 1994; Arnove and Christina 1998; Bano 2008a, 2008b; Ginsburg 1998; Klees 1998; Sutton and Arnove 2004). A third strand examines the role of international organizations and NGOs in producing transnational educational policy

(Moutsious 2009; Mundy and Murphy 2001; Robertson and Verger 2012). Less attention has been given to the interconnectedness of NGO advocacy and the concrete mechanisms that turn neoliberalism—a "theory of political economic practices" (Harvey 2007, 2)—into a way of being in the world (Feldman et al. 2016).

My research fills this gap in the literature by looking at how advocacy efforts in the refugee arena are mediated by the logic of the market. Located in an understudied refugee landscape of the Global South, my study contributes to a body of research on the circumstances that turn specific aspects of neoliberal practices into a means for NGO sustainability amidst high levels of government pressure. More specifically, it illustrates how NGOs navigate the contradictions that are generated when Correa Delgado's self-proclaimed socialism meets the neoliberal practices that loom large in the development world. Findings suggest that in response to the Ecuadorian government's attempts to politically control NGOs that serve people in refugee-like situations—including new regulations to oversee funding and highly demanding requirements regarding data collection (Instituto de Comunicación y Desarrollo 2014)—this hostile environment increased the competition among NGOs for the limited resources available from international donors. Not surprisingly, the drive to create sustainable programs in this constrained context has shaped the social advocacy efforts of NGOs. While national and international NGOs collaborated to expand the rights of migrants and act as one stronger policy advocate against the state, social advocacy was pursued in a competitive manner. In this arena, NGOs marketed their institutional identities and achievements to guarantee funds for survival.

In line with Kamat (2004), I argue that NGOs should be understood as extensions of the dominant neoliberal order. I further maintain that the implementation of a national development agenda, infused with the principles of Correa Delgado's political agenda, Socialism of the 21st Century, as described later in chapter, did not reduce the relevance of neoliberal practices as a means to guarantee NGO survival. On the contrary, the pressure exerted by Correa Delgado's government shoved NGOs further to resort to what Feldman et al. (2016, 3) characterize as "liquid advocacy," a changing and flexible form of advocacy that avoids disruptive political claims in order to secure funding and ensure organizational survival.

This research is significant because it draws attention to advocacy practices generally perceived within the NGO arena as innocuous. By linking debates about NGOs to a broader discussion of the mechanisms by which neoliberalism becomes reinvigorated, the chapter raises questions about the politics of "doing good" (Rodríguez-Gómez 2016a), a field we want to see as free from the political and economic interests that shape other spheres of our lives. This study interrogates organizational practices within a global development regime and the way that they serve to expand a model that is premised on the idea that human wellbeing can be achieved through entrepreneurial freedoms (Harvey 2007).

The Expansion and Contractions of Ecuador's Refugee Policies and Practices

In 2000, the Ecuadorian government called upon UNHCR to respond to the increasing influx of Colombian forced migrants. Until then, the *Comité Ecuménico Pro-Refugiados* (Ecumenical Committee for Refugees)—created in 1976 through an agreement between UNHCR and the Ecuadorian Episcopal Conference—had been overseeing refugee protection efforts in the country (Navas et al. n.d.). The expansion of UNHCR in Ecuador was intrinsically linked to the rising number of externally displaced people crossing the border from Colombia, and the tense bilateral relations between Colombia and Ecuador during the Andean Diplomatic Crisis. The crisis occurred in 2008 when the Colombian national military attacked a camp from the Revolutionary Armed Forces of Colombia (FARC) located over the Ecuadorian side of the border. Hand in hand with UNHCR, international NGOs arrived, and existing NGOs adapted their programs to target refugee populations. As in other areas of the Global South, such as Rwanda and Sierra Leone (see Betts et al. 2012; Vayrynen 2001), UNHCR Ecuador relied on NGOs not only to implement programs, but also to support social and policy advocacy efforts.

During my time in Ecuador (2013–2014), NGOs in the education sector witnessed their capacity to partake in the provision of education diminished. The desire of President Correa Delgado (2007–2017) to reverse the effects of "market reforms adopted at the end of the twentieth century" (Kent 2014, 421), and to reestablish authority and control over the educational system, translated into highly centralized policies and the corresponding expansion of the state (Baxter 2016). Aligned with Presidents Hugo Chávez in Venezuela and Evo Morales in Bolivia, Correa Delgado promised to turn away from neoliberalism and toward a new and more inclusionary society (Gustafson 2014). Criticizing "the long, cold, dark night of neoliberalism" (Correa Delgado quoted in Burbano de Lara 2015, 28). Correa Delgado instituted Socialism of the 21st Century, a political agenda that favored the reconcentration of power in the executive, and thus the expansion of state control over the distribution of resources. Correa Delgado's second national development plan, Governing to Deepen Change (*Gobernar para profundizar el cambio*), claims: "Our hands have not shaken to fight over the economic powers. The principal agent of neoliberal capitalism, the financial-banking capital, saw its capacity to exert political influence over governmental decisions largely reduced" (35 País 2013, 20).

Under previous governments, including the presidencies of Lucio Gutiérrez (2003–2005) and Alfredo Palacio (2005–2007), NGOs occupied a vacuum left by a failed state unable to fulfill the most basic functions, thus earning considerable space to operate. Under Correa Delgado's rule, NGOs were perceived as foreign carriers of neoliberal concepts, values, and practices, and therefore as oppositional forces to the president. His ideological combination of anti-neoliberalism, anti-imperialism, and nationalism explains the greater control his government exerted over NGOs.

Thus, it is not surprising to note that in contrast with the previous role of NGOs in Ecuador, during Correa Delgado's presidency, the Ministry of Education became the only education policymaker (Ortiz-Lemos 2013). This political move pushed national and international NGOs to the margins of educational governance, especially with regard to evaluation and monitoring efforts, and NGOs saw their capacity to influence the internal workings of the educational sector gradually reduced.

The struggle over access to education for migrant populations illustrates this shift. In 2003, four years before Correa Delgado was elected, the Education and Migration Working Group (EMWG) was created. It consisted of 114 organizations that included the UNHCR, ten national and international NGOs, and six migratory networks that had a presence in 14 provinces of Ecuador. The EMWG drafted what in 2006 became known as Ministerial Decree 455. Thanks to the actions of the EMWG, migrant children's access to education was established by law for the first time. Once the decree was passed, the EMWG started supervising its implementation. Relying on data collected during the monitoring process, the EMWG also took advantage of the principles of universal citizenship and universal access to education that were articulated in Ecuador's new constitution and drafted content for a new accord. Responding to the pressure of the EMWG in 2008, the Ministry of Education amended Decree 455 with Accord 337, which was more comprehensive and flexible and which guaranteed access to school even for children who were undocumented or who did not possess academic records (Rodríguez-Gómez 2016b). The following statement from an NGO director, during the interviews for this study, shows the capacity the EMWG had back then to define educational policy:

> Raul Vallejo was the minister at the time then. A group of us representatives from institutions sat down to talk to him, and to let him know what the problem areas were. He delegated the matter to an official and we found plenty of receptiveness at the time; let's say that [Decree] 455 came out of that same agreement. What did we do in between? [We] monitored and assessed the same process, and of course while monitoring the 455 we found that there were still many gaps that didn't allow getting the answer that we needed. Then we proposed once again the [Accord] 337 and we also found that same receptiveness.

This account shows NGOs' capacity to open and maintain a dialogue with those in power at the Ministry of Education in the first months of Correa Delgado's term. The use of the informal term *conversar* in Spanish gives a sense of horizontality and openness in describing the exchange between the NGOs and the minister. This statement shows that NGOs were policy makers and evaluators of educational policies.

However, after Accord 337 was passed in 2008, interviewees from my study described a divide arising between the governmental and NGO sectors. They reported not only the weakening of the EMWG—one said, "It's like the working group stood still for quite a while. . . . [There was] a year when it almost died, with a few little shoves, very little ones."—but also a distance between the governmental and the non-governmental spheres. With concern, one NGO employee told me about the difficulties he faced between 2010 and 2011 in doing his job as a political advocate in the chancellery:

Interviewee: Political advocacy was difficult back then.
Interviewer: What made it so difficult?
Interviewee: The change in the migratory policy in Ecuador. . . . The Ecuadorian government shut down contributions from civil society or it determined that it was authorized to maintain relations with international organizations like UNHCR and not with civil society organizations.

Surprisingly, a UNHCR representative reported: "We entered a very complicated period in 2010–2012 in our relationship with the state, because the door was closed." Even though participants could not articulate the reasons why the door had closed between the state and the non-governmental sector, the closing created a new set of obstacles that NGOs had to overcome to guarantee their sustainability.

Amidst an oil-fueled bonanza, Correa Delgado's government felt empowered to severely limit the role of international assistance in Ecuador, and to return to the state many of its traditional functions, including regulating citizens' access to basic rights. Under the premise of national sovereignty, Executive Decree No. 16, which was promulgated in June 2013, counteracted the expansion of NGOs by establishing a national secretariat to regulate the activities of national and international civic organizations, including the receipt of funds, both domestic and foreign. The secretariat had the power to control, and even dissolve, organizations critical of the government (Human Rights Watch [HRW] 2013; Ortiz-Lemos 2013). Uncertainties about the long-term sustainability of NGOs created an environment in which organizations were entangled by the competing demands of the state and donors, and their own desire to satisfy the needs of their participants. As my study findings show, these pressures shaped the advocacy practices employed by NGO staffers. In the marginal space permitted for NGO work by the state, NGOs retreated inwards by prioritizing the technical aspects of governance and by moving away from participation in major political debates, thus avoiding confrontations with political authorities. Even though interviewees recognized high-ranking officials' willingness to collaborate with NGOs, they also acknowledged that enthusiasm did not always translate into action. As the EMWG lost dynamism, the influence that NGOs had previously exerted over educational policy also got diluted. The state's effort to regain control implied a realignment of governmental and NGO sectors, creating a new and hostile environment for NGOs.

A Framework for Understanding Social Advocacy

"Non-governmental organization" (NGO) is an umbrella term for very diverse forms of associations, which include local, national, and international civil society organizations. In this chapter, I refer to all these organizations as NGOs because this is how my study interviewees referred to them. I define an NGO as a type of nonprofit organization that channels aid and delivers state-like services despite being institutionally separated from the state (Fisher 1997). In addition to their role as service providers, NGOs engage in some forms of advocacy, based on their relevant expertise, such as the representation of marginalized people to advance their rights. Advocacy covers a wide range of activities that often includes the collection and dissemination of data and high-profile technical advising for national governments, but advocacy also can comprise visible and public actions such as street-based public activities, demonstrations, and mass media campaigns to raise awareness.

While advocacy work entails the strategic production and dissemination of information, it also involves the translation and brokerage of key messages across diverse social actors, sectors, and levels. Jordan and Van Tuijl (1998, 1) define advocacy as the "righting of unequal power relations," but this is not always the case. The conceptualization of advocacy I use in this chapter emphasizes social advocacy as a means to mediate between powerful actors and the individuals and groups excluded from a system. I consider how, amidst the ascension of neoliberalism, the advocacy efforts of NGOs—of which mediation and brokerage are crucial elements—reconcile, for their own survival, organizational needs with the needs of those seeking international protection. In a context where NGOs are constrained by state control, and donors maintain a target-driven approach that implies increasing demands regarding an intervention's impact, budget allocations, and systems of accountability, the position of NGOs can be highly contradictory. Conditions characterized by a culture of heavy audit reduces NGOs' adaptability, preventing exploration of alternative strategies to respond to ever-changing contexts (Wallace 2009).

Neoliberalism is a slippery doctrine which, when applied to migration issues, normally implies the nexus of cheap labor, development, and the absence of citizenship (Chauvin et al. 2013; Levit 1999; Mossin Brønden 2012). Here, I apply certain dimensions of the neoliberal doctrine—in particular, practices such as competition for economic resources and the marketization of advocacy efforts—to explain the frames for action shaping the advocacy strategies implemented by the NGOs that serve individuals with refugee status. According to Ferguson and Gupta (2002), neoliberalism extends the logic of the market and the enterprise model to the daily workings of both government and NGOs. Given that neoliberalism is not a homogeneous concept (Feldman et al. 2016), to produce a more detailed examination of how neoliberalism operates on the ground, I rely on Ferguson's (2009) analysis of the difference between neoliberal doctrine and the regime of policies and practices associated with it. According to Ferguson, neoliberal doctrine calls for the deployment of market-based techniques outside the private sector. These techniques value the model of private enterprise, the elimination of taxes, and the deregulation of currency.

In tension with Mosse (2005), who highlights the demobilizing effects of development through marketization, Ferguson shows how organizations have the capacity to appropriate fragments of neoliberal doctrine through conscious "discursive and programmatic moves" (2009, 174). This differentiation allows us to understand the survival of neoliberal practices in realms that are not necessarily neoliberal per se, and how these practices reconfigure the relationships of NGOs with the individuals for whose rights they advocate. Drawing on Feldman et al. (2016), I maintain that when it comes to social advocacy, NGOs execute these programmatic moves through the adaptation of their programs and the marketization of their social advocacy initiatives. These programmatic moves seek to strengthen the economic position of NGOs in a competitive market through risk-averse behaviors and high levels of community participation. This combination masks the perpetuation of the neoliberal order with a sense of novelty and people's involvement, which allows NGOs to ensure their sustainability in a hostile environment.

An Analysis of Education Social Advocacy NGOs in Ecuador

To examine the role of NGOs as education stakeholders in the context of neoliberalism, the purpose of this research is to delve into the material conditions of NGOs working in the refugee-aid arena in Ecuador. This chapter relies on ethnographic work, more specifically semi-structured interviews with NGO and UNHCR employees, and participant observation in two international NGOs and one art exhibition.

Research Methods

Research was conducted in Quito because of the city's extensive network of NGOs and multilateral agencies, in contrast with other areas of Ecuador. Like Schmidt (2007), I conceive the interaction between NGOs and UNHCR as an interdependent and uneven set of relationships. Based on this premise, I selected organizations and participants to study through "purposeful network sampling" (LeCompte and Schensul 2010, 76). I first asked an initial contact to list her contacts at other organizations, then approached all of them for interviews, interviewed those who agreed, and then repeated the process until I had spoken with at least with one education staff member at each organization, thereby obtaining a sample of the larger network of NGO staff working in this field. The criterion for including NGO staff members was that they must be directly involved in education programs and for directors interviewed was that they must have oversight of educational programs but did not have to be directly working in education. Over a 13-month period between May 2013 and June 2014, I formally interviewed ten staff and three directors from seven NGOs. To understand how state and donors' demands shape NGOs' advocacy efforts, I asked staff members about their daily tasks, paying attention to the impediments they encountered to complete service provision and meet advocacy goals, and the strategies they implemented to overcome them. To document the other side of these relationships, I interviewed three UNHCR staff members, including Ecuador's UNHCR representative. These interviews provided inside information about the positions and practices of the United Nations representatives regarding refugee issues in Ecuador. All interviews were conducted and coded in Spanish; I translated the quotations included in this chapter into English. In order to capture patterns in NGO advocacy efforts, I coded the interview transcripts and daily notes to track the roles, relationships, and rules that defined advocacy efforts among NGOs.

To expand my understanding of the micro-social processes that shape social advocacy, and to pursue my research questions ethnographically, I attended the World Refugee Day in 2013 and 2014, and also volunteered for ten months at two of the international NGOs. As a volunteer, I not only observed, but also actively engaged in NGO tasks. I designed and implemented two educational projects for children and youth, which entailed recruiting participants who would benefit from the workshops, running the workshops on a weekly basis, and reporting back to my coordinators. In addition, in January 2014 I visited and analyzed the Flows art exhibition created by a local NGO with artistic artifacts produced by youth with refugee status, an experience I analyze later. The main purpose of the exhibition was to create a new language to represent issues related to youth and refuge in Ecuador (Fundación Museos de la Ciudad 2013). The exhibit, which lasted six months, offers a window into a form of social advocacy in the making.

To protect the confidentiality of the interviewees, I have used pseudonyms. Similarly, to protect the confidentiality all of the organizations included in this study, I refer to all of them as "NGOs." I found that the workplace experiences of NGO employees did not vary much, as their organizations relied on the same funders and served the same population, and thus I use the term NGO interchangeably.

Study Findings

Two main findings emerge from this study. First, pressure from the government and limited funding have translated into fewer opportunities for collaboration between NGOs and the state. These factors motivate NGOs to compete for public recognition and the limited economic resources available from international donors. Even though competition for resources between NGOs is not a neoliberal invention, precariousness and a growing audit culture has turned fundraising into the most obvious strategy for organizational survival. Second, the NGOs' competitive mindset translates into soft social advocacy efforts. The advocacy effort analyzed here illustrates a model of noncontroversial and risk-averse advocacy where those with refugee status are invited to advocate for themselves but face restrictions in participating in NGO decision making.

The NGOs' Endless Race for Financial Support

Lack of support from the state, as I aforementioned, along with the existence of limited private[2] and international funding, has engendered competitiveness among NGOs. Even though competition had always existed, since the 1980s, competition over acquisition of resources (including funding), recognition for accomplishments, and beneficiaries for their programs, has characterized the everyday workings of these organizations (Edwards and Hulme 1996; Klees 1998; Wallace 2009). NGOs have related to each other as competitors in a fierce and reduced market where everyone must struggle to survive. Thus, the possibilities for collaboration were limited by market-driven rationales.

In the refugee NGO arena in Ecuador, financial sustainability depends upon very few donors. Employees from six of the seven organizations that participated in this study indicated that their organizations shared the same funding sources: UNHCR; the Bureau of Population, Refugees and Migration (BPRM); and the European Community (ECHO). As the director of one NGO put it, "Unfortunately, the competitiveness that exists among institutions is due to the fact that in the end we are all competing for the same resources." A high-ranking official from UNHCR confirmed this assertion. From his point of view, two factors constrained the availability of funds: the absence of humanitarian work (in contrast with other UNHCR operations in camp settings where resources are more abundant) and the conditions for funding created by donors. Below, he explains how donor criteria for funding created competition between NGOs and even between NGOs and the UNHCR:

> There is some competition with regard to the resources and for the most part that depends on the donors. For example, there are donors that

---

2 Private sources of funding are limited in Ecuador. A study of the role of corporate social responsibility in the Andean region indicates that in Ecuador only 7 percent of the funding provided by corporations is channeled through NGOs. Corporations account for 42 percent of such social investments and churches the remaining 10 percents (Instituto de comunicación y desarrollo 2014).

understand that the actions of an international organization are different from the actions of civil society, so they maintain separate budget lines. . . . There are other donors like the European Community that place all of them in the same [category]. [If] you have to pass a migration and asylum bill, in fact, you have to compete against the Jesuits, [the] Hebrew International Aid Society, and even the UNHCR.

The race for resources came along with the need to align the missions and programs of NGOs with those of their donors to increase the NGOs' chances for funding. Anxiety around funding increased when NGOs that had established expertise in other areas shifted their mission toward refugee-related issues. As Blanca, who moved to Ecuador as a volunteer in 2009 and who in 2013 had a high-ranking position within her NGO, put it, "We have seen organizations that used to deal with coffee issues and nowadays work with youth. It is simple—there are some financing lines that guide the strategy [that funders will support], and from there organizations draft their projects and apply for funds." New donor agendas, therefore, resulted in the abandonment of particular populations not prioritized by donors and in the duplication of efforts across organizations, as multiple organizations scrambled to be considered for funding for donor-desired projects.

Following a larger and global trend toward emphasizing youth services in the international development agenda (see Péreznieto and Hamilton-Harding 2013), between 2013 and 2014 NGOs working to promote the education of those with refugee status prioritized youth (aged 12–18) over other populations, such as toddlers or children. As Blanca, explained, "The NGOs need to carry on projects, depend on donors; and at the present time world policy is focused on youths and adolescents. All the projects are aimed at youths and adolescents, including ours." In a context where the state did not fund NGOs and private donors invested little in NGOs, she found her NGO's portfolio of activities as too dependent upon donors' agendas and too similar to what other NGOs offered. For instance, because funding was oriented toward those under age 18, two different NGOs implemented moviemaking projects that focused on working with youth who had refugee status. One was called *Pipoca*, the Brazilian word for popcorn, and the other was named *Chulpicine*, which is a combination of the Ecuadorian word for dried corn kernels (*chulpi*) and the Spanish word for cinema (*cine*). To explain this duplication of activities, one NGO staff member borrowed the language of the market: "That a young person is able to decide whether to participate in program X or program Y is the normal logic. I can decide [for example] whether to attend the French course in a particular place or somewhere else." This approach naturalizes the logic of a competitive market where duplication is recast as the provision of choice for consumers, and NGO projects are expected to market themselves and compete with similar initiatives. Jairo, an NGO employee with more than three years of experience with youth, echoed this rationale:

> I haven't had to force young people or their family members to attend [my programs], which indicates that the activities we offer are attractive, interesting, and motivating. We cover their transportation costs, but we haven't come to the point of saying, "Well, if you are going to receive these resources you have to bring your children to the youth group," but that happens in other spaces.

Jairo's emphasis on the attractiveness of this project suggests the marketization of the activities provided by his NGO. Because attractiveness shapes levels of acceptance and consumption, it emerges as a key dimension of how competition for finite economic resources and market rationality, two of fundamental tenets of neoliberalism, enters the NGO domain.

NGOs competed not only for participants and funding, but also for reputation and visibility, two elements that strengthen their competitive position in scarce resource environments. During an interview, Juanita, an NGO staff member with more than five years of experience in refugee issues, stated clearly that the same rules applied to acquiring a good reputation and gaining access to money: "You need to demonstrate that you are working in order to obtain financing; as a result, economic resources and public recognition end up being one and the same." Similarly, concerned with her NGO's lack of participation in meetings organized by the UNHCR, the director of one NGO said, "We wanted to be part of the committees for durable solutions, and the response when I arrived was that we could not participate because we were not UNHCR operators." When asked why she cared so much about those committees, she clarified, "The issue of resources, unfortunately, is not everything, competitiveness is always present, it is competitiveness for recognition, for space, for a name, to be seen." Even though NGO employees referred beneficiaries to other NGOs when they were unable to satisfy their needs, NGOs did not have external incentives to cooperate. Oscar, a staff member with broad experience in education policy advocacy, summarized the obstacles to collaboration in the following terms:

> Cooperation spaces do not work, because there are no clear guidelines or objectives and the people [NGO employees] only show up because they have to. Also because "let's agree to work together" may mean that I have to surrender my territory; therefore, I'd rather play the fool.

The lack of collaboration meant that despite a tacit agreement in which some NGOs exercised control over certain areas of the city—note that Oscar says "my territory"—interviewees figured out that all organizations probably worked with the same population. This phenomenon was also observed by Schuller (2009) in Haiti, where NGOs "divide turf" (87).

These comments from NGO staff members show the hostile environment in which NGOs had to fight for funds; the fierceness of competition for different forms of resources pushed each NGO to stand out from the others.

In an environment where NGOs navigated the tensions created by a high degree of state control, NGOs applied economic rationales and marketization, two of the main practices associated with neoliberalism, to survive in such competitive environments.

Social Advocacy, Risk-Averse Behaviors, and Innovation by NGOs in Neoliberal Times

Here I situate my research findings on the work of NGOs in the context of Ecuador's Socialism of the 21st Century agenda by describing my impressions when I arrived to Quito to celebrate the World Refugee Day for the first time, and my close observations of Flows—an art exhibition organized by a local NGO. This exhibition is an example of NGO's social advocacy efforts to raise awareness around refugee issues.

As I planned my first trip to Ecuador in 2012, I learned that June 20th was World Refugee Day, and Quito would celebrate it on June 17th. It was a timely coincidence. I could not miss the opportunity to attend, and thanks to the generosity of an international NGO, just a few days after my arrival, I managed to secure a position as a volunteer for the day. My assigned responsibility was to ride a bicycle from downtown Quito to La Carolina—Quito's central park—with a group of children and youth identified by the organization as refugees. When the day arrived, under the auspices of the UNHCR, I received my blue t-shirt and cap with the white logo: "Live Together in Solidarity—World Refugee Day." After being introduced to each of the young people in my group, we rode to La Carolina. Once in the park, I identified the obvious signs of a commemorative event: a stage decorated with a large banner with institutional logos, loud sound speakers, and 15 vendor stands. Migrants selling food and Colombian handicraft businesses shared their stands with each other, whereas governmental and NGOs had one stand per organization—each decorated with institutional vinyl banners, and plastic tables covered with fliers, buttons, and brochures. This event setup shows an individualistic display of space and paraphernalia by NGO participating in the event. In contrast with policy advocacy activities where NGOs establish a coalition for negotiating with the government, participation in this World Refugee Day event demonstrates the logic behind the social advocacy work of NGOs, which was undertaken in isolation from their colleagues. NGO employees revealed a strategic mindset when they differentiated the strategies for advocating for policy reform from their segregated efforts to influence public opinion. As shown in this study, political advocacy was experienced as a collective undertaking where employees from different ranks across national and international organizations come together under the same purpose. Conversely, social advocacy was assumed to be a *solo mission* and a unique opportunity to expand each NGO's capital, specifically their reputation and brand.

The research I later conducted while embedded in two NGOs and one social advocacy effort, allowed recognition of social advocacy as an exercise of translation and brokerage between actors located at different levels of decision making. Through this lens, I analyzed an ongoing art exhibition called Flows, crafted in 2012 by a local NGO. The NGO was created back in 2000 to stimulate Ecuador's socioeconomic and sustainable development. Since 2004 the organization had been UNHCR's strategic partner in the implementation of entrepreneurship, and non-formal educational programs for those with refugee status. In contrast with other NGOs included in this study, this organization depended solely upon UNHCR resources and thus implied an urgent need to secure this relationship.

Flows, one of the activities implemented by this NGO started as a project with no funding. For reasons NGO staff members could not explain, the UNHCR stopped funding this NGO's community work in Quito's *barrios*. Weeks after the NGO ceased working there, those who had benefited from its presence called the NGO demanding its return. With limited funds, the NGO decided to come back to facilitate recreational spaces for youth, such as hip-hop and art workshops. Frustrated with the lack of financial support, an employee responsible for youth outreach, Jairo, invited UNHCR officials to witness the work he was doing in Quito's neighborhoods:

> I had to motivate them [UNHCR], so I tried to involve some people from UNHCR in the process. I invited them to the workshops we were doing with youth, working on getting them to comprehend the potential of the project. Basically, we were saying, "Listen, we know what we are doing, finance us."

Jairo declared this a win-win strategy: It allowed his NGO to gain recognition and support from UNHCR, and it allowed UNHCR personnel to get in touch with what was happening in the field. Aware of the role that visibility and recognition played in guaranteeing funding, Jairo's NGO decided to combine youth art workshops and social advocacy into one project: A multimodal art exhibit created in alliance with Quito's Center for Contemporary Art (CAC) that would help raise awareness on the situation of those with refugee status. The creative process behind Flows started in 2012 when the organization signed an alliance with the Foundation Museums of the City (*Fundación museos de la ciudad*) and the CAC. The purpose of this partnership was to collaborate on a "process of research, reflection and artistic production around the themes of refuge, territory and community from the perspective of the youth" (Fundación Museos de la Ciudad 2013). The exhibition aimed to "sensitize public opinion on issues related to refuge in Ecuador, [and] interrupt the [current] imaginary and adult-centric discourse on the subject" (Fundación Museos de la Ciudad 2013). The exhibition stayed open to the public at the CAC between June 2013 and May 2014, and later it traveled around Quito, more specifically in schools and community centers; it included a collection of artistic artifacts created by youths, and visual and oral testimonies of their life trajectories. The artifacts exhibited included videos, jewelry, silk screened prints, and photographs from hip-hop workshops and camps attended by Colombian and Ecuadorian youngsters. Alongside the artifacts from these workshops, the exhibition contained an organizational chart giving credit to the network of organizations and facilitators that contributed to Flows. Moreover it also presented testimonies from youth reflecting upon their experience, as expressed in these two comments:

> We liked sharing ideas, and together we wrote what we wanted, having someone to talk to, spend some time together. I learned many things; I was not only in silk screening. I used to go with Peter [rap teacher] or some classmates to other courses. I would have liked to continue there. I would not change anything about the project.
> I am very grateful to all those who helped us come through and I'm glad to have been part of [the workshop]. It was an experience that was magical for my life; it helped me discover talents I didn't know I had.

All these testimonies noted the youths' positive experiences in the NGO extracurricular activities. By emphasizing the participants' encouraging perceptions of the workshops, the testimonies served as evidence of the benefits generated by the project, working as a subtle but potentially powerful platform to "sell" the project among potential donors. This perspective aligns well with how Jairo conceptualized the exhibit:

> [We took advantage of] the media, the image of the CAC, because the exhibition was like raising a showcase, a platform, and a podium to present the issue [refuge] using a language that UNHCR would have never thought of. . . . I talked to the person who at the time was in charge of the auditorium: "Listen, this auditorium is used to present community processes" Why don't we do something? . . . and then we said, "well, we'll see," because to get to the presentation, in some way you have to have products to present. . . . But the workshops always yield products, so it goes more like going around collecting all the products, and analyzing what could be done with them. . . . Then we had to sit among the adults, the workshop facilitators, me, the people from the community work, and the CAC technicians . . . and then begin preparing the setup process"—Did youth participate? "It wasn't possible because those were technical issues, but in the end they saw everything and said "Wow."

By using artistic artifacts created by youth, and incorporating videos with their personal testimonies, the exhibition departed from popular perceptions of Colombians as disruptive and violent, and presented this population in a positive light. But what does this exhibition tell us about social advocacy at the time when Correa Delgado's socialism met neoliberal practices? Drawing from Ferguson (2009), I found that Flows is a programmatic effort within the neoliberal logic—a form of soft social advocacy—that results from the combination of risk-averse behaviors and creative innovation (Feldman et al. 2016). The exhibition not only raised awareness on the situation of young forced migrants, but also presented them as productive members of society; thus, the exhibition was risk averse because its creators chose to represent youth with refugee status in a way that that did not cause any discomfort or controversy among refugee stakeholders, in particular government representatives and potential donors. In other words, it did not challenge the structural violence that causes transnational forced migration, nor did it denounce the lack of state support for those in need of international protection—doing so would have hurt the NGO's already weak relationship with the government, and its ability to raise much-needed funds. In line with Hasenfeld and Garrow, I maintain that the preoccupation with marketability and sustainability dampens NGOs' motivation to challenge the state and even the UNHCR, "curtailing [their] historical mission to advocate and mobilize for social rights" (2012, 295). At the same time, the exhibit was an innovative example of social advocacy because, in the refugee-aid landscape, where contemporary art is perceived as an alien creative expression, the organization not only established a successful alliance with the CAC, but also created a new language for discussing issues related to refuge, one that departed significantly from the typical depiction of forced migrant youth as victims.

This exhibition represents a type of advocacy that combines the original social advocacy purpose of expanding human rights and raising awareness about the situation of those in need of international protection with the marketization of the experiences and artistic endeavors of young people. In this way, youth art became commodified with the purpose of raising additional resources to ensure the NGO's sustainability. The influence of neoliberalism becomes self-evident when we recognize how Jairo's narrative prioritizes the aesthetic dimension of the exhibition over youths' participation in the curating process. Aesthetics in this case matter, because they mediate the visibility and reputation of the exhibition and the NGO that produced it. To have a high-quality display was so important to Jairo and his team that instead of integrating youth in the process of curating the exposition, they preferred simply to show them the final result, thereby eliminating the opportunity for collaborative work between youth and adults. Jairo even implied that the youth did not have the capacity to comprehend the technical dimension of the work they had to accomplish. Moreover, his perception of the exhibition as "a showcase, a platform, and a podium" in conjunction with the organizational chart reveals the NGO's preoccupation with the visibility the exhibit could bring to its work. The inclusion of the NGO's name as part of the display showed how visibility, when produced through social advocacy, had as much value as other types of currency.

We also can see how the NGO passed the responsibility of representing refugees—advocating for them—to those with refugee status themselves. While the NGO curated the exhibition and planned the opening, workshop participants were held accountable for actually representing "the refugee experience". However, this form of participation kept youth away from the spaces where important decisions about their living conditions were made. Flows exemplified how under neoliberalism, NGOs create new forms of social advocacy that balance soft strategies to expand constituents' social rights with the means to guarantee their financial sustainability. Not surprisingly, before the exhibit's tenure at the CAC ended, Jairo's received good news: UNHCR would continue funding the project.

Conclusion

Based on data that I collected over a 13-month period, this chapter provides an ethnographic perspective on the extreme pressure that NGOs in the refugee landscape face in Quito. It shows how their work to promote migrants' access to education in Ecuador has led to their competing against each other in a context that brings together a contemporary form of socialism with neoliberal practices. As a result of the realignment between the governmental and the non-governmental spheres during Rafael Correa Delgado's presidency, the state closed its doors to collaboration between organizations and between organizations and the government. When the government exerted greater control over the non-governmental sphere, it created the conditions for NGOs to adopt marketization and competition strategies, two practices associated with neoliberalism, jostling among themselves to create programs that would appeal to outside donors. Drawing on a critical definition of neoliberalism, this chapter shows how the market force permeates the daily workings of the refugee-aid landscape, and how there is an ethos of economic calculation that attaches value to humanitarian work through its potential to engage in the domain of the market. Yet, I recognize that local advocacy efforts displayed by NGOs are not a unique response to the conditions imposed by the Correa Delgado regime. All NGOs included in this study, except for one, are international organizations that face similar constraints and design similar responses based on transnational scripts that stress the value of branding, competition, and systems of accountability, among other strategies, in a risk-averse environment.

By concentrating on the set of practices associated with neoliberal policies, such as the use of market-based techniques borrowed from the for-profit sector, including marketization and competition for resources, and focusing on some of the practices that turn neoliberalism into a repertoire for decision making, this chapter illustrates how competition among and between NGOs and their social advocacy activities are two spaces where humanitarianism intersects with certain aspects of neoliberalism. In a context characterized by intense resource competition coupled with lack of state support, my research findings show that, far from simple altruism, NGOs' social advocacy efforts have the potential to become rational strategies in the struggle for organizational survival. Through an analysis of Flows—an art exhibition seeking to raise awareness about the life trajectories of Colombian youth with refugee status—social advocacy emerges as a sophisticated enterprise, one that allows NGOs to value the model of private enterprise, by turning competition, marketization, and innovation into a catalog of survival strategies. Despite the fact that Flows used a new language to represent Colombian youth living in Quito, it did not allow these young people to take control of the exhibition or participate in broader decision-making processes. Indeed, one of the most important dilemmas of education advocacy within the context described by this chapter is how to denounce situations of inequality and empower rights holders—two of the main elements of social advocacy—without jeopardizing the competitive position of the organization *vis-à-vis* the state and the donor community. I therefore argue that forms of social advocacy that avoid risk hamper the potential of NGOs to address the unequal distribution of power or act as a force for democratization.

References

35 País. 2013. *Programa de Gobierno 2013–2017. Gobernar para profundizar el cambio. 35 Propuestas para el Socialismo del Buen vivir*. https://carlosviterigualinga.files.wordpress.com/2012/12/programa-de-gobierno-2013-20171.pdf

Archer, David. 1994. "The Changing Roles of Non-Governmental Organizations in the Field of Education (in the Context of Changing Relationships with the State)." *International Journal of Educational Development* 14: 223–232.

Arnove, Robert and Rachel Christina. 1998. "NGO-State Relations: An Argument in Favor of the State and Complementarity of Efforts." *Current Issues in Comparative Education* 1: 46–48.

Bano, Massoda. 2008a. "Non-Profit Education Providers Vis-á-Vis the Private Sector: Comparative Analysis of Non-Governmental Organizations and Traditional Voluntary Organizations in Pakistan." *Compare* 38: 471–482.

Bano, Massoda. 2008b. *Idara-Taleem-o-Aagahi's Partnerships with the Ministry of Education, Non Governmental Public Action Programme*. www.idd.bham.ac.uk/research/pdfs/Pakistan_ITA-Education.pdf

Baxter, Jorge. 2016. *Who Governs Educational Change? The Paradoxes of State Power and the Pursuit of Educational Reform in Post-Neoliberal Ecuador (2007–2015)*. PhD diss., International Education Policy, University of Maryland.

Betts, Alexander, Gil Loescher. and James Milner. 2012. *The Politics and Practice of Refugee Protection*. New York: Routledge.

Blum, Nicole. 2009. "Small NGO Schools in India: Implications for Access and Innovation." *Compare* 39: 235–248.

Burbano de Lara, Felipe. 2015. "Todo por la patria. Refundación y retorno del estado en las revoluciones bolivarianas." *Iconos, Revista de Ciencias Sociales* 52: 19–41. doi: 10.1714/iconos.52.2015.1670

Chauvin, Sébastien, Blanca Garcés-Mascareñas and Albert Kraler. 2013. "Employment and Migrant Deservingness." Special Issue: Migrant Legality and Employment in Europe. *International Migration* 51: 80–85. doi: 10.1111/imig.12123

Edwards, Michael and David Hulme. 1996. "Too Close for Comfort? The Impact of Official Aid on Nongovernmental Organizations." *World Development* 24: 961–973.

Feldman, Guy, Roni Strier and Michal Koreh. 2016. "Liquid Advocacy: Social Welfare in Neoliberal Times." *Journal of Social Welfare*. doi: 10.1111/ijsw.12250

Ferguson, James. 2009. "The Uses of Neoliberalism." *Antipode* 41(S1): 166–184.

Ferguson, James and Akhil Gupta. 2002. "Spatializing States: Toward an Ethnography of Neoliberal Governmentality." *American Ethnologist* 29: 981–1002.

Fisher, William. F. 1997. "Doing Good? The Politics and Antipolitics of NGO Practices." *Annual Review of Anthropology* 26: 439–464.

Fundación Museos de la Ciudad. 2013. *Derivas. Memoria y representaciones de refugio*. Quito, Ecuador: Fundación Museos de la Ciudad.

Ginsburg, Mark. 1998. "NGOs: What's in an Acronym?" *Current Issues in Comparative Education* 1: 29–34.

Gustafson, Bret. 2014. "Intercultural Bilingual Education in the Andes: Political Change, New Challenges and Future Directions." *The Education of Indigenous Citizens in Latin America*. ed. R. Cortina. Bristol, UK: Multilingual Matters.

Harvey, David. 2007. *A Brief History of Neoliberalism*. Oxford: Oxford University Press.

Hasenfeld, Yeheskel and Eve E. Garrow. 2012. "Nonprofit Human-Service Organizations, Social Rights, and Advocacy in a Neoliberal Welfare State." *Social Service Review* 86: 295–322.

Human Rights Watch (HRW). 2013. *Ecuador: Clampdown on Civil Society*. Decree's "Big Brother" Powers Undermine Groups' Independence. www.hrw.org/news/2013/08/12/ecuador-clampdown-civil-society

Instituto de Comunicación y Desarrollo. 2014. *Estudio regional sobre mecanismos de financiamiento de las organizaciones de la sociedad civil en América Latina*. http://mesadearticulacion.org/wp-content/uploads/2014/10/Informe-regional-mecanismos-final.pdf

Jordan, Lisa and Peter van Tuijl. 1998. *Political Responsibility in NGO Aadvocacy: Exploring Emerging Shapes of Global Democracy*. Global Policy Forum, Unpublished paper.

Kamat, Sangeeta. 2004. "The Privatisation of Public Interest: Theorizing NGO Discourses in a Neoliberal Era." *Review of International Political Economy* 11: 155–176.

Kent, Eaton. 2014. "The Centralism of 'Twenty-first-century Socialism': Recentralising Politics in Venezuela, Ecuador and Bolivia." *Journal of Latin American Studies* 47: 1130–1157. doi: 10.1177/0010414013488562

Klees, Steven. 1998. "NGOs: Progressive Force or Neo-Liberal Tool?" *Current Issues in Comparative Education* 1(1): 49–54.

LeCompte, Margaret. D., and Jean J. Schensul. 2010. *Designing & Conducting Ethnographic Research: An Introduction*. Lanham, MD: Altamira Press.

Levitt, Peggy. 1999. "Social Remittances: A Local-Level Migration-Driven Form of Cultural Diffusion." *International Migration Review* 32: 926–949.

Lewin, Keith. M., and Yusuf Sayed. 2005. *Non-Government Secondary Schooling in Sub-Saharan Africa: Exploring the Evidence in South Africa and Malawi*. http://

ia201113.eu.archive.org/tna/20061211103954/www.dfid.gov.uk/pubs/files/non-gov-2nd-schooling-africa.pdf
Mosse, David. 2005. *Cultivating Development: An Ethnography of Aid Policy and Practice.* London, UK: Pluto Press.
Mossin Brønden, Birgitte. 2012. "Migration and Development: The Flavour of the 2000s." *International Migration* 50: 2–7.
Moutsious, Stavros. 2009. "International Organisations and Transnational Education Policy." *Compare* 39: 469–481.
Mundy, Karen and Lynn Murphy. 2001. "Transnational Advocacy, Global Civil Society? Emerging Evidence from the Field of Education." *Comparative Education Review* 45: 85–126.
Navas, Albertina, José Francisco Sieber and Martin Gottwald. (n.d.) *La protección internacional de refugiados: El caso Ecuador: Perspectiva histórica 1976–2004.* Venezuela: ACNUR and Conferencia Episcopal Venezolana.
Nishimuko, Mikako. 2009. "The Role of Non-Governmental Organizations and Faith-Based Organizations in Achieving Education for All: The Case of Sierra Leone." *Compare* 39: 281–295.
Ortiz-Lemos, Andrés. 2013. *La sociedad civil ecuatoriana en el laberinto de la Revolución Ciudadana.* Quito: FLACSO.
Ospina, Oscar R., and Lucy Santacruz. 2011. *Refugiados urbanos en Ecuador: Estudios sobre los procesos de inserción urbana de la población colombiana refugiada, el caso de Quito y Guayaquil.* Quito: FLACSO.
Péreznieto, Paola and James Hamilton-Harding. 2013. *Investing in Youth in International Development Policy: Making the Case.* www.odi.org/sites/odi.org.uk/files/odi-assets/publications-opinion-files/8413.pdf
Robertson, Susan and Antoni Verger. 2012."The Rise of PPPs in Education: History and Conceptual Debates." *Public Private Partnerships in Education: New Actors and Modes of Governance in a Globalizing World.* ed. S. Robertson, K. Mundy, A. Verger and Fr. Mensahy. Cheltenham: Edward Elgar.
Rodríguez-Gómez, Diana. 2016a. *The Refugee Label: Mapping the Trajectories of Colombian Youth and Their Families through Educational Bureaucracies in Ecuador.* PhD diss., Columbia University, Teachers College.
Rodríguez-Gómez, Diana. 2016b. "When War Enters the Classroom: A Case Study on the Experiences of Youth on the Ecuador-Colombia Border." *(Re)Constructing Memory: Education, Identity and Conflict.* ed. J. Williams and M. Bellino. The Netherlands: Sense Publishers.
Rose, Pauline. 2007. NGO Provision of Basic Education: Alternative or Complementary Service Delivery to Support Access to the Excluded? *CREATE Pathways to Access.* Brighton: University of Sussex. www.create-rpc.org/pdf_documents/PTA3.pdf
Rose, Pauline. 2009a. "Non State Provision of Education: Evidence from Africa and Asia." *Compare* 39: 127–134.
Rose, Pauline. 2009b. "NGO Provision of Basic Education: Alternative or Complementary Service Delivery to Support Access to the Excluded?" *Compare* 39: 219–233.
Schmidt, Anna. 2007. " 'I Know What You're Doing': Reflexivity and Methods in Refugee Studies." *Refugee Survey Quarterly* 26: 82–99. doi: 10.1093/rsq/hdi0245
Schuller, Mark. 2009. "Gluing Globalization: NGOs as Intermediaries in Haiti." *POLAR: Political and Legal Anthropology Review* 32: 84–104. doi: 10.1111/j.1555-2934.2009.01025.x.

Sutton, Margaret and Robert Arnove. (2004). *Civil Society or Shadow State? State/ NGO Relations in Education*. Stamford, CT: Information Age.

United Nations High Commissioner for Refugees (UNHCR). 2017a. *Syria Emergency*. www.unhcr.org/syria-emergency.html

United Nations High Commissioner for Refugees (UNHCR). 2017b. *ACNUR en Ecuador*. www.acnur.org/t3/fileadmin/scripts/doc.php?file=t3/fileadmin/Documentos/RefugiadosAmericas/Ecuador/2015/ACNUR_Ecuador_2015_General_ES_Mayo_v2

Vayrynen, Raimo. 2001. "Funding Dilemmas in Refugee Assistance: Political Interests and Institutional Reforms in UNHCR." *IMR* 35: 143–167.

Wallace, Tina. 2009. "NGO Dilemmas: Trojan Horses for Global Neoliberal?" *Socialist Register* 40: 202–219.

6 The Implications of Education Advocacy for Civil Society Organizations

Constanza Lafuente

This chapter extends our understanding of advocacy as presented in the Introduction of this volume by discussing the implications of education advocacy with respect to the policies and practices of civil society organizations[1] (CSOs). Not intended to be a step-by-step list of possible advocacy strategies and activities, it instead examines six organizational practices that support education advocacy by applying concepts from relevant literature and using applicable examples from the case studies presented in earlier chapters of this volume. Aligning organizational missions with activities and resources, establishing cooperative relations with other actors in their organizational environments, and diversifying income sources are examples of these practices.

To examine the practices that support education advocacy, I draw on concepts presented in studies investigating advocacy activities and practices of non-governmental (NGOs) and nonprofit organizations. Although comparative education research is increasingly exploring these concepts, studies have not examined the organizational implications of adopting education advocacy. The research concerning NGOs, for example, explores the practices, roles, and behaviors of civil society organizations involved in international development in two types of organizations: those with headquarters in the North that operate programs overseas (Northern NGOs) and those with their headquarters in the Global South. Research studies on nonprofit organizations tackle domestic issues in the United States.

While some of the challenges faced by nonprofits in the United States differ from those that CSOs experience in the Global South, there are nevertheless a number of similarities between them. Hence, incorporating concepts from both fields of research enhances the understanding of organizational

---

1 The Introduction defined CSOs as the organizational infrastructure of civil society (Lewis 2007). The landscape of CSOs includes organizations with various ideologies, structures, and resources. Grassroots organizations, cooperatives, Southern and Northern non-governmental organizations, mutual aid associations, religious schools, community-based organizations, transnational advocacy networks, and student organizations are examples of types of CSOs.

practices and offers richer perspectives on the workings of the groups involved in education advocacy. For example, although CSOs in the Global South are predominantly located in contexts of highly unequal resources and, in some cases, unstable political environments, northern nonprofits and southern CSOs combine social goals with private structures, and are shaped by their need to garner resources in a competitive environment, work with volunteers and community members, and provide services or advocacy for underrepresented populations.

The audiences for this chapter are practitioners (such as leaders, managers, directors, and volunteers) of CSOs, operating locally or nationally, that are already engaged in advocacy work; students and practitioners who wish to learn about the internal workings of organizations involved in education advocacy in order to incorporate such advocacy into CSOs that have been engaged in other activities; and practitioners who would like to reflect on their current practices.

The first section of this chapter defines social and political advocacy, with special attention to education advocacy, based on the framework presented in this volume's Introduction. The second introduces the strategy elements of education advocacy. The third section discusses the organizational environments of CSOs and illustrates how advocacy strategies influence an organization's environment. The fourth section analyzes six practices that support education advocacy. The final section provides guidance on how to incorporate education advocacy effectively into an organizations' programs and on how to generate the funds needed to do so.

The Concept and Operationalization of Advocacy

The advocacy approach specifically targeted to the modification of public and private actors' decisions or practices on behalf of groups that receive a low-quality education builds upon the more general definition of education advocacy previously presented in the Introduction of this volume. Advocacy can be the main strategic organizational approach or an ancillary approach that complements the services provided by a CSO. Within the field of international education development in particular, advocacy includes a wide repertoire of actions that seek to eliminate education disparities and their root causes, as seen by the cases included in this book (see Table I.1 in this volume's Introduction) for an examination of the various advocacy emphases existing in the education field.

Service delivery of education differs from education advocacy in that the former does not seek to change policies and institutions, as is the case with CSOs that provides a variety of supports or distribute goods, but conforms to existing governmental policies (Minkoff 1999). Other CSOs may design innovative services and use the evidence of the impact of their services to influence the design and implementation of education policy. In education, service provision—which can be implemented with or without government funding—seeks to improve the quality and reach of various types of education assistance for underserved groups. Services by CSOs may include non-formal education provided outside of government schools, such tutoring or arts education; private provision of education,[2] of which parochial schools are examples; or programs that seek to improve the quality of public education, such as the *Escuela Nueva* Foundation (see Chapter 4 in this volume).

Building upon Jenkins' (2006) conceptualization of advocacy, social and political advocacy are two main emphases in education advocacy. Through social advocacy, CSOs devoted to education seek to shape the opinions, practices, and agendas of private organizations and individuals, including other education CSOs. Through political advocacy, CSOs target policy makers to influence decisions and education policies through the various policy stages, such as agenda setting, the enactment of policy and regulations, implementation and monitoring. Organizations may also seek to influence decisions of the judiciary so that it guarantees and protects the right to education.

The Strategy Elements of Education Advocacy

This section presents a view of strategy as both a concept and a set of practices that shed light on the workings of organizations that adopt education advocacy. It builds upon research in the nonprofit and non-governmental organizational fields (Austin et al. 2006; Lewis 2007; Phills 2005; Sheehan 2010) that highlights the following elements related to development and implementation of a strategy: (a) an organizational mission and vision that orient social goals and actions; (b) programs, activities, internal processes, and resources that interact in helpful ways to fulfill the organizational mission and vision; and (c) explanations of how programs, processes, and resources interact with external actors, such as donors, other CSOs, and schools, and with other environmental factors, such as the political environment, to bring about change in education. All these elements are embedded in the organizational culture defined through the values and norms of the members of an organization (Law 2016), including boards of directors, staff, volunteers, and members.

As seen in Figure 6.1, a CSO's mission, values, and culture are at the core of its strategies, with social goals in the next innermost circle. Both circles are aligned to advocacy and service delivery-related programs. Organizational activities and resources are designed to support those programs.

---

2 *Fe y Alegria*, the biggest network of nonprofit and Catholic schools in Latin America, founded by the Jesuit order and financed through public, international cooperation and private funding is an example of an education organization that provides education services for the poor.

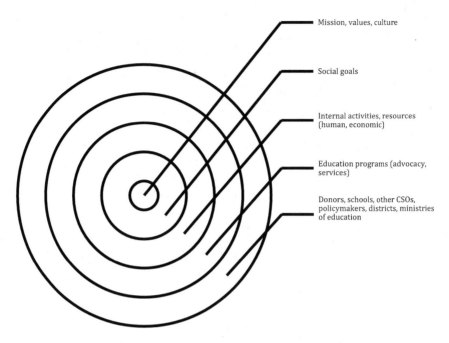

Figure 6.1 CSO's Strategy Elements

Strategies define how programs, activities, processes, and resources interact internally and with groups and individuals in organizational environments. In the case of CSOs devoted to education, groups and individuals may include schools, policy makers, donors, districts, other CSOs, and education-related organizational networks (local, national, and transnational).

As seen in this conceptualization, the notion that CSOs can influence their external environments underlies the concept of strategy (Fowler 2016). This notion reflects the holistic perspective of CSOs that encompasses the relations between these groups and their environments (Anheier 2005); it assumes that although the environment influences the workings of organizations, it is also true that the organization, reciprocally, influences the environment.

The Organizational Environments of Education CSOs

Here I complement the conceptualization of strategy elements by presenting a framework for understanding the organizational environments of education CSOs. This framework demonstrates how advocacy strategies influence an organization's environment. All organizations are located within broader environments consisting of other organizations, institutions, and events that facilitate or hinder access to key resources (Pfeffer and Salancik 2003). The work of Lewis (2007) on the strategic NGO management framework allows the assumption that CSOs can control some elements of their organizational environments, influence other elements, and grasp but not easily modify other elements (see Figure 6.2). The framework also highlights the role of education practitioners in understanding and interpreting their environments to better inform their advocacy strategies (Lewis 2007; Ronalds 2010).

Education CSOs can control and manage elements in the first level of their environments, including their staff, values, education programs, and budgets, although control is never complete because organizations are open systems (Scott 1998) and certain actors, such as donors, also seek to influence aspects of their practices. At the second environmental level, CSOs can seek to influence education policy, allocation of resources earmarked for public education, decision-making processes, agendas, citizens' attitudes toward public education, and the practices of other CSOs. They generally cannot, however, control their donors' agendas, networks, or other CSOs through their advocacy strategies. As I put forward in the next section, to influence such environments, organizations can seek to pursue strategies on their own, or they can collaborate with individuals or other groups in civil society, which requires having a full understanding of the policies, practices, and goals of all relevant groups in their organizational environments.

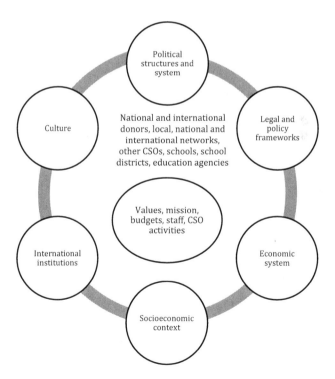

Figure 6.2 The Organizational Environments of Education CSOs, adapted from Lewis (2007)

At the third level of CSOs' organizational environments, which includes socioeconomic contexts, economic and political systems, cultural institutions, and international organizations, education CSOs neither control nor influence these components, although alliances with social movements (see Chapter 3 in this volume), or transnational and national coalitions (Eickelberg 2012; Mundy 2012; Mundy and Murphy 2001) may advance changes at this third level.

This framework has several implications for education advocacy practitioners. First, it shows where advocacy strategies are included in the organization's environment. It also illustrates the importance of assessing the particular environments of a CSO, because factors such as location or field of intervention in education (i.e., early childhood, primary, secondary schools) can modify the extent to which certain aspects of organizational environments are more or less relevant for their strategies. Its size may also shape the extent to which a CSO can exert influence, with bigger and resource-rich organizations in a better position to influence its environments (Almog-Bar and Schmid 2014). Elements at each level may also change across time because environments are dynamic (Lewis 2007). Moreover, the fluid and changing nature of environments highlights the importance of organizational capacity to act strategically by proactively capitalizing on political and economic opportunities that arise.

Practices That Support Education Advocacy

Based on relevant existing research, I have identified six practices that support education advocacy: (1) developing cooperative relations with other groups in the organizational environments of CSOs; (2) designing organizational missions in support of education advocacy, and aligning programs and accountability with them; (3) combining service delivery with advocacy through scaling-up; (4) acquiring the technical skills and resources that are required to support advocacy; (5) diversifying resource generation strategies to enhance advocacy sustainability; and (6) obtaining organizational leadership support. Each is described below.

It is worth noting that a large number of factors determine the extent to which any or all of these six practices can be successfully incorporated into organization. The factors include size, levels of participation, location, strategies, structures, governance, policy fields of operation, ideologies, and the ratio of volunteers to paid staff. In Latin America, in particular, the landscape of possible organizations is diverse, as are the factors that define them.

Developing Cooperative Relations With Other Organizations to Support Education Advocacy

Based on the conceptualization of CSO environments, the first practice—establishing cooperative relations with other groups in CSOs' organizational environments—supports education advocacy by highlighting the role of practitioners in identifying actors who are relevant to increasing advocacy engagement. Practitioners' holistic views of their environments support development of such relationships because they enable the identification of groups with complementary missions, purposes, and strategies at the local, national, or transnational level.

The challenges for large-scale change in education are so daunting that CSOs cannot affect change on their own, such as reframing education as a human right. Thus, based on an understanding of their own strengths and limitations, they may seek to cooperate with other groups. Chapter 3 in this volume describes how CSOs can collaborate with student movements. Indeed, the chapter's authors, Cristián Bellei, Cristian Cabalin, and Victor Orellana, show that at times CSOs are not the main agents in affecting change in public education. CSOs certainly can, however, support the involvement of other groups, with the goal of enabling a social movement to exert influence in the broader macro environment greater than that of a CSO acting on its own. Indeed, Chapter 3 shows that Chilean CSOs supported the mobilization of student movements by providing technical assistance, which was based on their own research activities and organizational capabilities. Chilean student movements (e.g., the Penguin Revolution lead by high school students in 2006 and the Chilean Winter lead by university students in 2011) were powerful actors in challenging the marketization of the Chilean education system through which education was reduced to a technical activity and students were beneficiaries or consumers of education services. The student movements reframed the right to education as including more than just access, and the quality of educational content and processes was considered for the first time. Here, student movements led to social protest, while CSOs supported their mobilization and demands through their own activities. The support provided by CSOs was based on their own organizational missions, purposes, and capabilities (in particular research capacities) for they provided policy analysis support, which student movements used to substantiate their demands and proposals to legislators and policy makers.

Designing Organizational Missions in Support of Education Advocacy

Mission statements, oftentimes complemented by visions that express a desired future state, are the main blueprints for CSO performance and accountability. The mission of CSOs, unlike that of businesses, is to pursue social goals, not profits. This is the starting point for CSO education advocacy. Three issues that are relevant for education advocacy practitioners, are described below: mission statements that enable education advocacy, links between missions and accountability, and alignment between mission and an organization's programs.

Enabling Education Advocacy Through Mission Statements

Mission statements suggest some of the ways CSOs frame social and education issues, along with their own positions. Obviously, some statements support the adoption of certain types of education advocacy, while others discourage it. Mission statements may use terms that encourage the adoption of advocacy-related work, thereby becoming the starting point for education advocacy (Bass et al. 2014). Conversely, mission statements that discourage advocacy are those that highlight assistance and services, and support conformity with education policies. Other statements will support the simultaneous adoption of multiple strategy orientations, namely both advocacy and service delivery.

Below I analyze the mission statements of two organizations included in this volume—*Mexicanos Primero* (see Chapter 1) and *Escuela Nueva* Foundation (see Chapter 4)—and discuss how both statements support advocacy in different ways but still are conducive to advocacy. *Mexicanos Primero's* mission, as provided on its website (*Mexicanos Primero* 2017)—"to encourage an understanding of the shared responsibility around national priorities of education"—is an example of a mission statement conducive to advocacy as a core purpose, and it suggests that public education is a common responsibility of social actors. *Escuela Nueva's* mission, indicated in its website—"to contribute to improve the quality, relevance and efficiency of education by rethinking the way we learn and promoting active, cooperative and personalized learning" (*Escuela Nueva* 2017)—emphasizes the organization's education service model. As explained in Chapter 4, *Escuela Nueva* Foundation's education advocacy is exclusively linked to the scaling-up of its service model, an education program that encourages participatory, self-paced and multigrade learning in rural areas with participation of the communities in schools. Its vision complements its mission statement: "By 2018 we want to be a global technical reference for active, cooperative and personalized learning based on the *Escuela Nueva* model; and we want to lead a global movement to improve the lives of the underserved through our educational model centered on the learner" (*Escuela Nueva* 2017). Hence, the organization's advocacy is aligned with its mission and vision since the organization seeks to influence education policy and implementation by providing expertise to policy makers for the mainstreaming and scaling-up of its education service model.

Linking Mission and Accountability

Mission statements also serve as guides for performance (Sheehan 2010) and accountability (Jagadananda and Brown 2010), which, as explained in Chapter 2 in this volume, can indicate how CSOs report to authorities and stakeholders, and are held responsible for their outcomes (Edwards and Hulme 1996).

Defining organizational performance is probably one of the most challenging activities for CSOs. It can even become a contested activity for practitioners because organizations do not pursue profits, and internal and external stakeholders—the board, staff, members, and the individuals affected by organizational actions—may have different perspectives on what performance means. Moreover, deciding which groups and individuals take part in such discussions can result in conflicts within organizations and external stakeholders (Roberts et al. 2005).

Although discussing what performance means for all groups can significantly enhance the relevance and legitimacy of an organization's mission and actions—where legitimacy is defined through the "assumption that organizational activities are desirable, proper, or appropriate within some socially constructed system of norms, values, beliefs, and definitions" (Suchman 1995, 574)—oftentimes, practitioners are discouraged from engaging in such discussions. As discussed in Chapter 2, *Mexicanos Primero's* efforts to engage and account to teachers and parents would result in tremendous impact to the organization—in particular on their legitimacy, organizational learning, and strengthening ties with citizens. However, involving these groups entails challenges and tensions, for sometimes CSOs concentrate more on their own definition of effectiveness and overlook the perspectives of the individuals they are seeking to help, primarily those who receive a low-quality education. Practitioners tend to be under pressure to be accountable to donors or with internal stakeholders exclusively rather than to beneficiaries or the groups affected by the organization's actions.

In CSOs, *how* they function is as important as *what* they do (Edwards and Fowler 2002; Lewis 2007). Thus practitioners can invest time in learning from other organizations by exploring different ways in which those organizations have incorporated stakeholders in their discussions around mission and performance, in particular the groups affected by the organization's actions. Such activity requires resources (time, staff), but it will increase the relevance and legitimacy of CSOs as education advocates and also clarify what performance means for different groups as they learn about the perspectives of those they seek to serve or whose rights they want to advance. Thus, missions and performance should not only reflect internal values and *raisons d'être*, but also respond to needs and interests of target communities.

Aligning Mission and Programs

Aligning education programs and activities with mission statements and values is a key practice that supports education advocacy so that all organizational capabilities and resources are directed toward the activities that are key to mission fulfillment. When they are not properly aligned, practitioners may need to assess whether programs should continue or be revised so that they adequately reflect the values, purpose, and desired role of the organization in education, and also the rights of the communities that the organization seeks to advance. While practitioners may adjust regularly their strategies to produce incremental enhancements—strategy refinement supports continuous improvement (Ronalds 2010)—processes that include reviewing the mission and that may lead to broader changes in the overall strategy are undertaken less often because organizational values are permanent, and missions should not change as often as activities (Phills 2005).

Reflecting continuously on how programs align to their mission is a crucial activity for CSO education practitioners since departures from the organization's mission distracts an organization from its main purposes. *Mission creep* is defined as actions taken that move organizations away from their original purposes (Fowler 2010). Mission creep occurs, for example, when organizations adopt or modify programs to accommodate to donors' preferences in order to obtain new economic resources.

The quest for organizational sustainability and survival may lead organizations to expand programs into new intervention areas or incorporate new target populations that lie outside the boundaries of their organizational missions. In Chapter 5 of this volume, Diana Rodríguez-Gómez demonstrates that some practitioners aligned missions to donors' purposes as concerns about available funding increased in the CSO community. Her case study of CSOs in the refugee sector in Ecuador shows that one organization with operations in the coffee sector diversified its mission and activities by incorporating programs for youth with refugee status. She also recounts that CSOs working in the early childhood education field modified their activities as donors started prioritizing funding for youth and refugee issues. As she argues, practitioners in this case study sought to market the social advocacy component of their organizations, and one organization in particular, to increase chances for funding, suggesting that advocacy is not a heroic activity, but rather a practice embedded in organizations with that need to survive within an environment of limited resources. These findings highlight the need for practitioners to continuously reflect on their organizational strategy so that programs align to mission, and ultimately, capacities and resources are invested in the activities that are critical to the organizational mission.

Combining Service Delivery With Advocacy Through Scaling-Up

This section explains the practice of combining service delivery with advocacy. Blending both approaches is one way to facilitate CSOs' advocacy because organizations can use the evidence of "what works in the field" to influence the design and implementation of education policies.

CSOs' efforts to scale up their programs constitute one way to combine service delivery with advocacy. CSOs may first expand the reach of their programs across sites and build evidence on the effectiveness of their education interventions; they then can seek to modify the design and implementation of public education policies. CSOs may also pilot a program in one location and use the evidence of the program's effectiveness to scale it up with the government as a partner, or a private donors, so that it can reach multiple geographical areas. Organizations may also scale up their programs on their own, or partnership with other CSOs to extended their reach. The mainstreaming and scaling-up approaches described here are part of proactive strategies that seek large-scale change in education; they differ from passive program expansion where an organization increases the reach of its programs because there are funds available, not because influencing the design and provision of government-funded education programs is the main purpose of its strategy.

Escuela Nueva Foundation's *Circulos de Aprendizaje*, a program described in Chapter 4, combined service delivery with advocacy through mainstreaming and scaling-up. The case study shows that the Ministry in Colombia approached the organization with the intention of adopting its *Circulos de Aprendizaje's* program. Yet, the organization already had a tradition of scaling up its programs in partnership with the Ministry, with the purpose of supporting its education advocacy strategy of shaping the design and implementation of education policy. In the 1980s, with funding from the World Bank, the *Escuela Nueva* model had been standardized into a kit and scaled up to multiple rural regions across Colombia (McEwan 1998).

Below, using *Escuela Nueva* Foundation's *Circulos de Aprendizaje* case as an example, I discuss how the scaling-up approach supports education advocacy and some of the main components in mainstreaming and scaling-up processes.

Mainstreaming and Scaling-Up Education Programs

Through scaling-up strategies, practitioners use the evidence of the impact of their education services to replicate their programs in collaboration with other CSOs, international organizations, Ministries of Education, or regional districts. During the 1990s, as donors validated this practice by providing funding for it, and as more CSO practitioners envisioned it as a way through which they could increase their reach and social impact, scaling-up became a catchphrase (Edwards and Hulme 1992; Schnell and Brinkerhoff 2010; Uvin et al. 2000) and the new science of change in the field of development (Ditcher 1999). Practitioners also saw scaling-up approaches as opportunities to support their advocacy efforts to influence the design and implementation of education policy. In particular, scaling-up approaches could build on the operational and service-related expertise of CSOs to influence education policy.

There are various definitions of scaling-up: some highlight organizational quantitative growth (more coverage, more beneficiaries, bigger budgets, more geographic locations) and others highlight the scaling-up of social impact (the increased impact of programs through social franchising and influencing policy despite no increase in the size of the organization's operations). I define the term broadly here, invoking the concept of replication (Schnell and Brinkerhoff 2010) to generate widespread change in education. Replication, as Carter and Currie-Alder explain, brings "more quality benefits to more people over a wider geographical area more quickly, more equitably and more lastingly" (2006, 129). Mainstreaming education services by incorporating them into the government's programs is one of the

140 *Constanza Lafuente*

ways in which replication may occur. As depicted in Figure 6.3, I propose an understanding of scaling-up as an approach that supports political advocacy (Jenkins 2006), because it assists CSOs with service-related expertise in the process of influencing governmental decision makers that make decisions over the design and implementation of education policies.

Key elements in the mainstreaming and scaling-up of education programs, described below and illustrated with examples from the *Círculos de Aprendizaje* case, enable influencing the design and implementation of education policy.

As shown in Figure 6.4, practitioners can consider a number of elements for their mainstreaming processes, defined through the incorporation of the

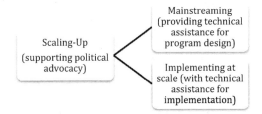

Figure 6.3 Scaling-Up to Promote Advocacy

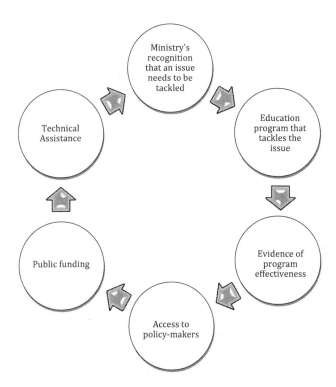

Figure 6.4 Elements That Support Mainstreaming of an Education Program

design of a CSO's program into a government program, including (a) a Ministry of Education's recognition that an education issue needs to be tackled; (b) a CSO's education program that seeks to resolve the issue; (c) evidence of the education program's effectiveness in tackling the issue (through impact evaluations, pilots, implementation research); (d) access to policy makers who make program and funding decisions; (e) public funding;[3] and (e) technical assistance provision to policy makers on program design.

After mainstreaming an education program, a government will often decide to scale it up to multiple regions, and CSOs usually provide technical assistance for this process (see Figure 6.5). To take education programs to scale by partnering with governments in a way that is sustainable and effective for participating schools and local communities, practitioners may also

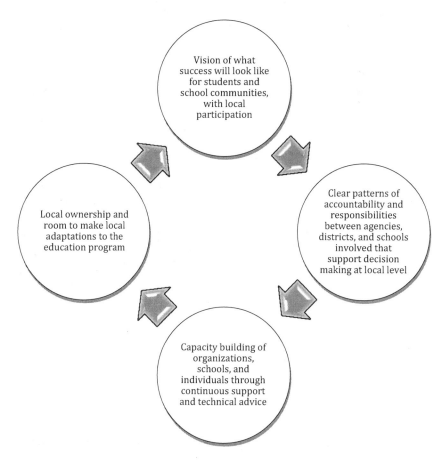

Figure 6.5 Elements That Support Scaling-Up of an Education Program

---

3 Although the focus here is to analyze scaling-up processes with governments as partners, when no government funding is available, governments and CSOs may also scale up programs through private or cooperation funds.

consider how to implement the following elements: (a) articulate a vision with local communities that defines success through quality, including what such quality could look like for students, classrooms, and schools, and the main conditions for achieving quality; (b) ensure a clear organizational design structure that assigns clear patterns of responsibility and accountability between the various agencies, districts, and schools involved, and that supports collaboration and decision making at the local level; (c) build the capacity of all participating organizations, groups, schools, and individuals by providing continuous support and technical advice; and (d) create local ownership, as evidenced through the participation of local schools and adaptations made at the local level to meet needs and cultural practices of participating districts, communities, and schools.

The case of *Escuela Nueva* Foundation's *Círculos de Aprendizaje* illustrates how the elements that support mainstreaming also enable taking the implementation of an education service to scale (see Chapter 4 for an analysis of the challenges of, and lessons learned from, the *Círculos de Aprendizaje's* program).

The previous operational successes of this organization through its *Escuela Nueva* model went hand in hand with a real need on the part of the Colombian Ministry of Education to tackle the issue of internally displaced children. Displaced children were defined as those who were out of the educational system for at least six months; their numbers had increased substantially during the early 2000s due to the spiraling of guerrilla and paramilitary groups. Thus, the Ministry decided to fund the implementation of a program to tackle this issue, and *Escuela Nueva* Foundation already had an intervention model in place—the *Escuela Nueva* model—that had previously been evaluated as an effective intervention. An evaluation conducted in 1997 by UNESCO's (United Nations Educational, Scientific and Cultural Organization) Latin American Laboratory for the Assessment of Quality in Education (LLECE) had provided evidence on the effectiveness of *Escuela Nueva's* interventions in improving students' learning in math and language (Casassus et al. 2000). Moreover, the United States Agency for International Development (USAID) had funded the design of a new program, *Círculos de Aprendizaje*, which was implemented as a pilot program in 2003 and 2004. *Escuela Nueva's* founder and CEO had direct access to the Ministry of Education who, after visiting the site where the *Círculos de Aprendizaje* program was being piloted, thought that the organization's programs could be instrumental in restoring the right to education for displaced children. The Ministry allocated public resources to the mainstreaming of the program and assumed responsibility for taking the program to scale across Colombia.

Findings from research on *Escuela Nueva* Foundation's *Círculos de Aprendizaje* also attest to the importance of the four elements that can support taking programs to scale, depicted in Figure 6. First, as findings from the case study, presented in Chapter 4 show, when the program was taken to scale, there was a lack of vision about what success would look like in terms of increased educational quality for students, modifications in the pedagogical practices in classrooms to support student reintegration, and changes in the schools that would align with the program's overall purpose of reintegrating out-of-school children. An emphasis on quantitative expansion, as opposed to quality, translated into a lack of attention to teaching and learning and pedagogical practices in schools. Consequently, the organizational and project conditions needed to achieve success were not delineated. Second, the organizational design was complex, and members of the regional teams struggled individually with their responsibilities. Moreover, the relationships between the various actors involved in the implementation at scale were strongly based on control and supervision, instead of ongoing support and guidance. Third, the communication between the national program operators and the regional teams who coordinated the work of the secretariats of education and the schools (*Institutiones Educativas Madre* [IEM]) was infrequent. Comprehension of the program by the regional teams was partial, which restricted their chances of providing technical assistance and support to develop capacities of the secretariats of education and IEMs. Finally, local communities and schools lacked ownership, as evident in the lack of local participation in decisions around program quality and implementation. For the most part, local participants had to comply with decisions rather than adapt the program to their particular contexts.

The lessons the *Escuela Nueva* Foundation learned through the scaling-up of the *Círculos de Aprendizaje* experience, show the importance of clarifying the patterns of responsibility among all parties involved, and of setting expectations for quality. As these lessons suggest, organizational reflection activities on project development and implementation are crucial so that practitioners can refine and improve their scaling-up approaches and are better prepared to better influence the design and implementation of education policies.

Although some policy makers, researchers, and practitioners praise the scaling-up of CSO services as a way to ensure the sustainable replication of CSOs interventions, others instead criticize the lack of local participation in the scaling-up of CSOs services and interventions. Findings in Chapter 4 suggest that scaling-up is not a one-size-fits-all practice in the quest to influence education policy design and implementation. As CSO practitioners continue to test various scaling-up models, reflecting on said experiences and using those learnings to inform their future scaling-up approaches will hopefully engage organizations in cycles of organizational learning and improvement.

Acquiring Technical Skills and Resources for Advocacy

This section examines how practitioners plan for the human and material resources they will need to support their advocacy activities as they design their approaches. Such planning highlights the importance of having the internal capacities and resources to support these activities and programs.

Practitioners need to consider if their organizations have the staff in general, and staff with the necessary technical skills specifically, and the organizational capabilities to support their education advocacy models. It would not be realistic to expect that CSOs—whether medium size or small—have all the skills to support multiple types of education advocacy simultaneously (Bass et al. 2014). Hence as an organization reflects about the possible addition of an advocacy component, it must consider its currently available capacities, capacities that could be further developed internally, and capacities that other CSOs and organizational networks in their organizational environments could provide.

While larger CSOs will tend to be in better organizational condition to pursue multiple advocacy activities and strategic emphases simultaneously, since they are likely to have more economic resources and specialized staff (Child and Grønbjerg 2007), most organizations have the capabilities, skills, and resources to adopt specific advocacy activities because each activity requires unique technical skills and resources. For example, some organizations will be better equipped with the necessary resources and capabilities to focus on social advocacy through public awareness campaigns or outreach to the media and citizens, whereas others will be better prepared to engage in political advocacy.

Two activities that are components of advocacy strategies, described below, illustrate the requirement for specific capabilities, staff, and resources. As seen in Chapters 1 and 2, describing *Mexicanos Primero's* strategies, and in Chapter 4, covering *Escuela Nueva*'s scaling-up efforts, these components feed into an organization's political and social advocacy activities that seek to improve public education: NGO journalism (McPherson 2014) and research.

Through NGO journalism, practitioners generate news content and various types of information for multiple groups (Powers 2015), and they also obtain media coverage that they use to support their advocacy campaigns (McPherson 2014). Campaigns to raise awareness on issues pertaining to public education and relations with the press are integral to *Mexicanos Primero's* advocacy strategy. Over time, the organization became a specialized source of information for journalists on education-related issues in Mexico. As the communications director of *Mexicanos Primero* explained, the organization "act(s) promptly upon consideration of everyday events," and has a team dedicated to analyzing current events, news, and activities. The organization has a staff of six full-time employees with a rigorous research background, in addition to volunteers who support the research team; they provide relevant information about education events to journalists; organize media-related events; and focus on specific media including newspapers, television, and digital media.

Research, the second advocacy activity provided here as an example, generates rich information that the organization uses for writing proposals and framing and assessing problems in education. Two research types support education advocacy, each with different organizational implications in terms of the human resources and technical skills required to conduct the research. The first, applied research, is incorporated into the organization's structure through its own research unit, and is seen in the case of *Mexicanos Primero*. The second, implementation research and evaluation, is conducted by third parties that assist education CSOs in validating their intervention programs. It is exemplified by *Escuela Nueva* Foundation's *Circulos de Aprendizaje*, where CSO practitioners used external research to support the mainstreaming of its education program. Whereas the first research type is an internal activity in which the CSO allocates specialized personnel and resources to conduct ongoing research, the second is conducted by third parties and does not require CSO personnel, although it may demand allocation of additional economic resources, which may be provided by external donors interested in testing the efficacy of an education intervention.

When practitioners determine that the organization does not have the capabilities and resources to develop certain desired activities, they may consider cooperating with other groups with complementary missions, purposes, and strategies that might provide the needed resources. Although CSOs often work independently, practitioners may also join local or regional organizational networks in civil society to encourage cooperation with other CSOs, and facilitate the sharing of costs and technical capacity to support their education advocacy strategies.

Diversifying Resource Generation Strategies

Organizations that adopt advocacy strategies can benefit from diversifying their sources of income, increasing their domestic funding and resources that are not attached to project implementation. As indicated throughout this chapter, advocacy is a broad concept that can comprise many different activities, such as providing technical assistance for policy design, lobbying legislators, and monitoring policy outcomes. Some activities, such as providing technical assistance and capacity building, are more tangible for donors and thus are easier to fund than other advocacy activities, such as lobbying legislators or implementing education campaigns, which do not produce measurable outputs and outcomes. Thus, challenges associated with the funding of education advocacy are linked to the particular approaches that organizations adopt.

Escuela Nueva's funding illustrates some of the challenges that a well-established education CSO faces. Its main activity is influencing the design and implementation of education policy by combining services and advocacy work. During its 30-year history, *Escuela Nueva* has been able to maintain a steady flow of funds that allowed the organization to test, refine, and scale up its programs at the national level. As reported by the Foundation's financial reports (*Escuela Nueva Volvamos a la Gente* 2016), the vast majority of the resources raised in 2016 (US$ 1,896,180) were attached to implementation of its programs and services, including curriculum design and training, and technical assistance to local school districts and public school teachers. As indicated above, the organization's advocacy is tied to the spread and scaling-up of its education programs. Thus, technical assistance contracts with the government for the implementation of *Escuela Nueva's* various programs provide a majority of its revenue (65 percent), followed by the funds from national and international foundations (15 percent) and national and international organizations (e.g., Inter-American Development Bank and private universities) (14 percent). Only a small portion of its revenue (3 percent) is unrestricted and provided by individuals.

Having a stable and ever-increasing flow of unrestricted funds so that the organization can fund research and development of new education models and programs is one of the goals for *Escuela Nueva's* fundraising activities. Creating new programs and innovating require having funds that the organization can invest freely. However, such activities do not produce quantifiable outputs, which donors tend to prioritize over those that cannot be quantified.

Professionalizing its fundraising team and building its reputation at the local level to attract individual and unrestricted donations are also goals for *Escuela Nueva* because a philanthropic culture is not well established in Colombia. Indeed, mobilizing domestic individual donations is difficult. To increase unrestricted funds, the organization is also expanding its fundraising efforts to the United States with the hope that it can attract unrestricted individual donations from its expat community abroad and other individuals residing in the United States. Moreover, having donors that commit social investments to more than a one-year cycle is another goal that the organization has set for itself since one-year-cycle grants jeopardize the sustainability of the projects and force the organization to engage in continuous donor solicitation and activities that absorb significant organizational efforts.

As the *Escuela Nueva* example illustrates, diversifying sources of income and increasing individual donations are key to increasing sustainability and innovation. Through the diversification of funding mechanisms and revenue sources, practitioners can enhance their sustainability. Moreover, diversifying sources of income may not only increase sustainability, but also strengthen the legitimacy of a CSO (Parks 2008), as organizations with diverse sources do not depend on one donor.

Funding trends in Latin America also highlight the relevance of diversifying CSO resource generation strategies to increase financial sustainability because the region overall is not a priority in terms of development aid. Latin America as a whole receives less official development assistance aid (ODA) than least developed countries (Pousadela and Cruz 2016). With the exception of Haiti,[4] Latin American countries are considered lower, upper, and higher middle-income economies (World Bank 2016).[5]

---

[4] Haiti is the least developed country in Latin America (World Bank 2016).
[5] El Salvador, Guatemala, Honduras, and Nicaragua are lower middle-income economies (World Bank 2016).

Moreover, regional donors fund education services rather than activities that do not produce quantifiable outcomes, such as research and development, campaigning, or networking with public officials and legislators. In addition, international funds are decreasing in Latin America overall, funding that in part was allocated to advocacy activities, although regions such as the Andean and Central American countries still receive high portions of international funding (*Instituto de Comunicación y Desarrollo* 2014).

Education is one of the most important targets for grants, donations, and social investments in Latin America, but in general, most funds are allocated to services or infrastructure projects. For example, a recent study by Hauser Center for Civil Society in Latin America, reported that education is the field to which wealthy individuals direct most of their donations (Johnson et al. 2015). Another study on the corporate social investments of *multilatinas* in education—Latin American companies with their headquarters in that region that have scaled up beyond their national markets—shows that two-thirds of corporate social investments are made to education through corporate foundations or firms' corporate budgets (Van Fleet and Sanchez Zinny 2012). Despite the importance that Latin American donors assign to education, gifts generally prioritize service delivery or infrastructure projects; financial support for education advocacy, including advocacy that seeks policy reform, is limited.

For example, a study on the role of *multilatinas* in education illustrates that most firms make small donations by funding local schools (77 percent) and local nonprofits (77 percent), and that a majority of investments fund teacher training (64 percent) and generally do not fund programs that seek policy reform or advocacy (Van Fleet and Sanchez Zinny 2012). Advocacy-related issues, such as governance reform (9 percent), grassroots and social movements (5 percent), and policies and planning (1 percent), receive little support from *multilatinas*. Similarly, another study about the funding strategies of CSOs in Latin America (*Instituto de Comunicación y Desarrollo* 2014) reported that organizations pursuing advocacy strategies tend to have less access to domestic funding than do service delivery CSOs because the latter have increased their chances of accessing public funds as service contractors attracting most of the government funds available (Pousadela and Cruz 2016).

To reduce the extent to which education advocacy organizations rely on external grants, depending on their main activities, CSOs will have to explore new ways of generating revenue. Building their reputations, seeking to increase individual funding, both nationally and internationally, and exploring new mechanisms, such as crowd funding and social impact bonds that monetize future savings to governments (Salamon et al. 2014) might be effective additions to the current traditional grant-seeking strategies.

Obtaining Organizational Leadership Support

The last education advocacy practice concerns the key roles of CSO leadership and governance in supporting advocacy strategies, with the caveat that this discussion may not be entirely applicable to those CSOs whose decision-making structures are based entirely on collective decision making (Roberts et al. 2005). Leadership structures depend on the specific organization and context (Phills 2005), and organizational leadership looks very different across various education organizations. Some CSOs have collective decision making or even rotate positions (Roberts et al. 2005). In membership-based CSOs members (individuals or organizations) elect the organization's leadership (Smith 2010), and in CSOs without members, the board of directors (bodies whose members elect themselves) appoint a director to be responsible for the management of the organization.

Although the literature on nonprofit organizations presents executive directors and boards as having two separate functions, with positions occupied by different individuals (the board is responsible for strategic direction, and directors are responsible for the overall management of organizations), a study on boards in Latin American CSOs (Berger 2010) found that both functions often overlap.

Irrespective of how such leadership and governance look across education CSOs, leaders are key to education advocacy because they can push for the formalization of advocacy approaches in the strategy of an organization, providing key resources and support. Further, in general, they are key in creating the conditions and practices that are conducive to advocacy, such as crafting, approving, or evaluating CSO organizational missions and strategies, and providing the necessary direction, resources, and support to enable advocacy. CSO leaders can also strengthen the legitimacy of CSOs as education advocates based on their personal expertise in education, and they can support alliances with key actors in organizational environments (other CSOs, academics, schools, unions, networks) that provide the crucial resources and support to increase advocacy engagement. Moreover, wherever possible, they can facilitate access to decision makers to influence policy design or implementation. Without such leadership support, CSOs would face significant internal constraints in implementing advocacy activities sustainably, and advocacy might end up becoming an activity that even if well executed may not be properly incorporated into the mission and strategy of the organization.

Conclusions

This chapter examined education advocacy from the perspective of the organizational strategies and the internal operations of CSOs, acknowledging that robust organizations, guided by the right missions and values, are in a better condition to advance improvements in education for all than organizations without clearly articulated strategies and alignment between resources and capacities.

As competition for material support increases in the field of international educational development, CSO practitioners will need to have the skills to align their resources and capabilities with organizational missions. In addition, increasingly, they will need to strategically assess, interpret, and develop opportunities offered by their organizational environments in ways that increase their advocacy engagement, effectiveness, and sustainability. Doing so will require practitioners to be realistic about how their capabilities align with their advocacy strategies and about the opportunities and constraints offered by the environment. These are topics that have traditionally received little attention by researchers in comparative education. This chapter has sought to offer education practitioners new insights into all aspects of the process of developing and implementing education advocacy as an organizational activity. The hope is that the analysis of case studies and recommendations based on the experiences of CSOs presented here will assist them in evaluating and possibly refining their daily practices and strategic decision making, and thereby increase the effectiveness of their advocacy.

References

Almog-Bar, Michal and Hillel Schmid. 2014. "Advocacy Activities of Nonprofit Human Service Organizations: A Critical Review." *Nonprofit and Voluntary Sector Quarterly* 43(1): 11–35.

Anheier, Helmut K. 2005. *Nonprofit Organizations: Theory, Management, Policy*. London, UK: Routledge.

Austin, James, Roberto Gutierrez, Enrique Oligastri and Ezequiel Reficco. 2006. "Effective Management of Social Enterprises. Lessons from Businesses, and Civil Society Organizations in Iberoamerica." Social Enterprise Knowledge Network (SEKN). Cambridge, MA: Harvard University Press.

Bass, Gary, Alan Abramson and Emily Dewey. 2014. "Effective Advocacy Lessons for Nonprofit Leaders from Research and Practice." In *Nonprofits and Advocacy*. ed. R.Pekkanen, S. R. Smith and Y. Tsujinaka. Baltimore: Johns Hopkins University Press.

Berger, Gabriel. 2010. "Boards." In *International Encyclopedia Of Civil Society*. ed. H. Anheier and S.Toepler. New York: Springer.

Carter, Simon and Bruce Currie-Alder. 2006. "Scaling-Up Natural Resource Management: Insights from Research in Latin America." *Development in Practice* 16(2): 128–140.

Casassus, Juan, Sandra Cusato, Juan Enrique Froemel and Juan Carlos Palafox. 2000. "Primer Estudio Internacional Comparativo sobre Lenguaje, Matemática y Factores Asociados, para Alumnos del Tercer y Cuarto Grado de la Educación Básica". *LLECE—UNESCO. Laboratorio Latinoamericano de Evaluación de la Calidad de la Educación*. Santiago de Chile: UNESCO. http://escuelanueva.org/portal1/images/PDF/Evaluaciones/13_unesco_esp.pdf

Child, Curtis D., and Kirsten A. Grønbjerg. 2007. "Nonprofit Advocacy Organizations: Their Characteristics and Activities." *Social Science Quarterly* 88(1): 259–281.

Ditcher, Thomas. 1999. "Globalization and its Effects on NGOs. Efflorescence or a Blurring of Roles and Relevance?" *Nonprofit and Voluntary Sector Quarterly* 28(38): 38–58.

Edwards, Michael and Alan Fowler, A. 2002. *The Earthscan Reader on NGO Management.* 1st ed. London: Earthscan.
Edwards, Michael and David Hulme. eds. (1992). *Making a Difference. NGOs and Development in a Changing World.* London: Earthscan.
Edwards, Michael and David Hulme. 1996. "Too Close for Comfort? The Impact of Official Aid on Nongovernmental Organizations." *World Development* 24(6): 961–973.
Eickelberg, Anja. 2012. "Framing, Fighting and Coalition Building: The Learnings and Teachings of the Brazilian Campaign for the Right to Education." In *Campaigning for "Education for All": Histories, Strategies and Outcomes of Transnational Social Movements in Education.* ed. A. Verger and M. Novelli. Rotterdam: Sense.
Escuela Nueva Volvamos a la Gente. 2016. *Cifras Economicas 2014–2016.* Bogota, Colombia.
Escuela Nueva Volvamos a la Gente. 2017. "Mission and Vision." http://escuelanueva.org/portal1/en/who-we-are/mision-and-vision.html.
Fowler, Alan. 2010. "Options, Strategies and Trade Offs in Resource Mobilization." In *NGO Management, The Earthscan Companion.* ed. A. Fowler and C. Malunga. London: Earthscan.
Fowler, Alan. 2016. "Non-Governmental Development Organizations' Sustainability, Partnership, and Resourcing: Futuristic Reflections on a Problematic Trialogue." *Development in Practice* 26(5): 569–579.
Instituto de Comunicación y Desarrollo. 2014. "Estudio Regional sobre Mecanismos de Financiamiento de las Organziaciones de la Sociedad Civil en America Latina." *Mesa de Articulación de Plataformas Nacionales y Redes Regionales de América Latina y el Caribe.* http://mesadearticulacion.org/wp-content/uploads/2014/10/Estudio-Mecanismo-Financiamiento.pdf
Jagadananda and David Brown. 2010. "Civil Society Legitimacy and Accountability: Issues and Challenges." In *NGO Management, The Earthscan Companion.* ed. A. Fowler and C. Malunga. London: Earthscan.
Jenkins, Craig. 2006. "Nonprofit Organizations and Political Advocacy." In *The Nonprofit Sector: A Research Handbook.* ed. R. Steinberg and W. Powell. New Haven: Yale University Press.
Johnson, Paula, Christine Letts and Colleen Kelly. 2015. *From Prosperity to Purpose: Perspectives on Philanthropy and Social Investment among Wealthy Individuals in Latin America.* Zurich and Cambridge: UBS Philanthropy Advisory and Hauser Institute. https://cpl.hks.harvard.edu/publications/prosperity-purpose-new-research-philanthropy-latin-america
Law, Jonathan. 2016. *A Dictionary of Business and Management.* New York: Oxford University Press.
Lewis, David. 2007. *The Management of Non-Governmental Development Organizations.* London, UK: Routledge.
McEwan, Patrick. 1998. "The Effectiveness of Multigrade Schools in Colombia." *International Journal of Educational Development* 18(6): 435–198.
McPherson, Ella. 2014. "Advocacy Organizations' Evaluation of Social Media Information for NGO Journalism." *American Behavioral Scientist* 59(1): 124–148.
Mexicanos Primero. 2017. "Nuestra Misión." www.mexicanosprimero.org/index.php/mexicanos-primero/lo-que-hacemos

Minkoff, Debra. 1999. "Bending with the Wind: Strategic Change and Adaptation by Women's and Racial Minority Organizations." *American Journal of Sociology* 104(6): 1666–1703.

Mundy, Karen. 2012. "The Global Campaign for Education and the Realization of 'Education For All.'" In *Campaigning for "Education for All": Histories, Strategies and Outcomes of Transnational Social Movements in Education*. ed. Antoni Verger and Mario Novelli. Rotterdam: Sense.

Mundy, Karen and Lynn Murphy. 2001. "Transnational Advocacy, Global Civil Society? Emerging Evidence from the Field of Education." *Comparative Education Review* 45(1): 85–126.

Parks, Thomas. 2008. "The Rise and Fall of Donor Funding for Advocacy NGOs: Understanding the Impact." *Development in Practice* 18(2): 213–222.

Pfeffer, Jeffrey and Gerald Salancik. 2003. *The External Control of Organizations: A Resource Dependence Perspective*. 2nd ed. New York: Harper and Row.

Phills, James A. 2005. *Integrating Mission and Strategy for Nonprofit Organizations*. New York: Oxford University Press.

Pousadela, Inés M., and Anabel Cruz. 2016. "The Sustainability of Latin American CSOs: Historical Patterns and New Funding Sources." *Development in Practice* 26(5): 606–618.

Powers, Matthew. 2015. "Contemporary NGO-Journalist Relations: Reviewing and Evaluating an Emergent Area of Research." *Sociology Compass* 9(6): 427–437.

Roberts, Susan, John Paul Jones and Oliver Fröling. 2005. "NGOs and the Globalization of Managerialism: A Research Framework." *World Development* 33(11): 1845–1864.

Ronalds, Paul. 2010. "The Change Challenge: Achieving Transformational Organizational Change in International NGOs." In *NGO Management, The Earthscan Companion*. ed. A. Fowler and C. Malunga. London: Earthscan.

Salamon, L., Stephanie Geller and Wojciech Sokolowski. 2014. *Navigating the Future: Making Headway on Sustainability for Social Accountability Organizations*. Working Paper 2. Washington, DC: Global Partnership for Social Accountability

Schnell, Sabina and Derick Brinkerhoff. 2010. "Replicability and Scaling Up." In *International Encyclopedia Of Civil Society*. ed. H. Anheier and S. Toepler. New York: Springer.

Scott, W. Richard. 1998. *Organizations, Rational, Natural and Open Systems*. Upper Saddle River, NJ: Prentice Hall.

Sheehan, Robert M. 2010. *Mission Impact*. Hoboken, NJ: Wiley.

Smith, David Horton. 2010. "Grassroots Associations." In *The International Encyclopedia of Civil Society*. ed. H. Anheier and S. Toepler. New York: Springer.

Suchman, Mark. 1995. "Managing Legitimacy: Strategic and Institutional Approaches." *American Management Review* 20(3): 571–610.

Uvin, Peter, Pankaj Jain and David Brown. 2000. "Think Large and Act Small: Toward a New Paradigm for NGO Scaling Up." *World Development* 28(8): 1409–1419.

Van Fleet, Justin and Gabriel Sanchez Zinny. 2012. *Corporate Social Investments in Education in Latin America and the Caribbean: Mapping the Magnitude of Multilaterals Private Dollars for Public Good*. Working Paper 5 (August). Washington, DC: Center for Universal Education at Brookings. www.brookings.edu/wp-content/uploads/2016/06/08-investment-latin-america-van-fleet.pdf

World Bank, 2016. *Country Classification Table*. Washington, DC: Center for Universal Education at Brookings. http://go.worldbank.org/47F97HK2P0

World Bank Group, 2014. *Johns Hopkins Center for Civil Society Studies*. http://gpsaknowledge.org/wp-content/uploads/2014/12/Navigating-the-Future-Making-Headway-on-Sustainability-for-SAcc-Organizations-GPSA-Working-Paper.pdf

Contributors

Cristián Bellei is an associate researcher at the Center for Advanced Research in Education, and an associate professor in the Sociology Department, both at the University of Chile. His research areas include educational policy, school effectiveness, and school improvement. He has published extensively about Chilean education in numerous academic journals. Among his most recent publications is "The Choice of School as a Sociological Phenomenon. A Review of the Literature" (2017, *Revista Brasileira de Educação* with Victor Orellana, Sebastian Caviedes, and Mariana Contreras). A sociologist from the University of Chile, he holds a doctorate in education and a master's degree in international education policy from Harvard University.

Cristian Cabalin is an assistant professor at the Institute of Communication and Image and an associate researcher at the Center for Advanced Research in Education, both at the University of Chile. He is also a researcher at the Universidad Central de Chile. As a journalist, he has been a contributor to the electronic newspaper *El Mostrador*; the magazine *Rocinante*; and the newspapers *The Guardian* (England) and *Milenio's* (Mexico). A recent publication is "The Mediatization of Educational Policies in Chile: The Role of the Media in a Neoliberal Education Field" (in *News Media and the Neoliberal Privatization of Education*, ed. Zane Wubbena, Derek Ford, and Brad Porfilio, 2016, Information Age Publishing). He holds a doctoral degree in educational policy studies from the University of Illinois at Urbana-Champaign and a master's degree in anthropology from the University of Chile.

Regina Cortina is a professor of education in the Department of International and Transcultural Studies at Teachers College, Columbia University. Her most recent book, *Indigenous Education Policy, Equity, and Intercultural Understanding in Latin America* (2017, Palgrave) is a comparative study of policies designed to increase the educational opportunities of Indigenous students, protect their rights to an education inclusive of their cultures and languages, and improve their education outcomes. *The Education of Indigenous Citizens in Latin America*

(2014, Multilingual Matters), her previous book, examines unprecedented changes in education across Latin America that resulted from the endorsement of Indigenous peoples' rights through the development of intercultural and bilingual education. Professor Cortina is president-elect of the Comparative and International Education Society (CIES) and will become president in March 2018.

Constanza Lafuente works at Carnegie Corporation of New York's Education Program, where she supports the Teaching and Learning to Advance Learning portfolio. There, she manages grantee relations and supports and informs program strategy. Before that, she worked in teacher professional development for New York City's Pre-K for All at Bank Street College of Education's Education Center. She is also an Adjunct Assistant Professor at Teachers College, Columbia University, where she teaches the graduate course, Non-Governmental Organizations in International Educational Development. Dr. Lafuente's professional experience includes early childhood and primary education quality improvement projects in New York and Argentina. She obtained a Ph.D. in comparative and international education with a concentration in political science from Teachers College, Columbia University, in 2010, with a Fulbright scholarship. She also holds a M.Sc. in public policy in Latin America from the University of Oxford.

Víctor Orellana is an adjunct professor at the Department of Sociology of the University of Chile and a researcher at the Center for Advanced Research in Education, also at the University of Chile. He also serves as the director of the "Commitment for a New Education" initiative at the *Nodo XXI* Foundation. One of his recent publications is "Why to Choose a Semi-Private School? Middle Social Class Sectors and School Choice in a Market System" (2016, *Estudios Pedagogicos* 43(3), with Manuel Canales and Cristián Bellei). His research interests include the sociology of education, especially education and social structure. A sociologist, he holds a master's degree in social science from the University of Chile.

Diana Rodríguez-Gómez is an assistant professor in the School of Education at Universidad de Los Andes, in her hometown of Bogotá, Colombia. Her academic and teaching interests gravitate around the intersections of violence and education in Latin America. Through an ethnographic approach that includes visual methods, her research focuses on the social processes that link global and national policies with classroom practices in contexts affected by violence. Her dissertation recently received an Honorable Mention from the Council on Anthropology and Education, and the Best Dissertation Award from the Comparative and International Education Society's Latin American Special Interest Group. She holds an Ed.D. in international educational development with an emphasis on human rights and peace education from Teachers College, Columbia University.

Laura María Vega-Chaparro works at *Fundación Escuela Nueva*, coordinating its community of practice, which brings together partners and stakeholders with the aim of creating new knowledge regarding active, cooperative, and personalized learning based on the *Escuela Nueva* model. Her professional experience includes working in high-poverty, violent, and vulnerability contexts; and collaborating with teachers, public school communities, and governmental, foreign aid, and civil society organizations. She has worked with the Colombian Ministry of Education and the *País Libre* Foundation. Her research interests include the study of citizenship and education, human rights, and peace education. A Colombian psychologist, she holds an Ed.D. in international educational development from Teachers College, Columbia University, with a concentration on peace education.

Index

accountability, 11–12, 38, 46–47, 49–51, 56, 59–61, 104, 114, 134–36, 141–42; external dimension of, 8, 43–49, 51; internal dimension of, 8, 43–49, 51–52, 58, 60; practices, 8–9, 42–46, 48, 50, 58–60
advocacy, 1, 4, 6, 10–12, 14–15, 23–24, 40, 42, 44–45, 108, 114, 129–32, 136, 138–39, 147–49; actions, repertoire, 12, 29, 40, 130; education advocacy, 4, 6, 11, 13–15, 44, 60, 125, 129–31, 134–37, 139, 144–45, 147–48; political, 1, 6–7, 10, 24–25, 86, 88, 106, 113, 120, 130–31, 140, 144, 150; social, 7, ·10, 11–12, 21, 24–26, 28–29, 33, 108–10, 114–16, 120–21, 123–25, 131, 144; strategies, 6–8, 12, 18, 24–25, 39–40, 48, 88, 114, 129, 133–34, 144–45, 147, 149
Anheier, H. 46, 47, 132
Archer, D. 4, 109
Arnove, R. 4, 109, 160

Bartlett, L. 86, 105
Bendell, J. 42, 46, 51, 59
Brown, L. D., Jagadananda 45–46, 48–49
Brunner, J. J. 67–68, 81
Bureau of Population, Refugees and Migration (BPRM), 117

Carnoy, M. 67
Chile, 2, 9, 12, 16, 63–69, 71–73, 77, 79–84, 107, 149, 153–54; education, 5, 9, 64–65, 68, 71–81, 153; higher education, 67–69, 71, 72, 78–80; market oriented reforms, 3, 65, 66–68; Penguin Revolution, 9, 65, 68, 72, 74–76, 135; student movements, 63–64, 69, 72–76, 79–80, 83, 135; Winter, 9, 65, 135
Christina, R. 109
civil society, 1, 4–5, 8–10, 12–13, 15–18, 22, 37–41, 43, 60–62, 64, 128–29, 133, 145, 149, 151
Clark, J. 24
Coburn, C. 94, 104
Cohen, J, Arato, A. 4, 18
Colbert, V. 86–87, 90, 104
Colombia, 85–105; displaced children, 10, 85, 88, 90; Forces of Colombia (FARC), 109, 111; Ministry of Education 85, 87, 90–96, 99–105
Comparative Education, 4, 16–17, 39, 41, 44, 61, 63, 105, 125–26, 149
Correa Delgado, R. 109–13, 123–24

Ebrahim, A. 43, 45–46, 49–51, 56, 61
Economic Commission for Latin America and the Caribbean (ECLAC), 3, 16
Ecuador, 2–3, 10, 41, 108–125; Flows, 109, 116, 120, 121, 122, 123, 124, 125; Ministry of Education, 112; refugee, 108–111, 114, 116–19, 121–124, 138
Education for All (EFA), 1–2, 4, 11, 16–17, 22, 34, 40–41, 45, 58, 61–62, 71, 87, 127, 148, 150–51
education policies, 4, 6–8, 10, 20–21, 23, 25–26, 75, 77, 79–80, 130–31, 133, 136, 138–40, 143, 145
Edwards, M. 42, 44, 46, 48, 85, 105, 117, 136, 137, 139
Eickelberg, A. 44, 134

Index

Elacqua, G. 70
Escuela Nueva, 10, 85, 87–88, 105–7, 136, 142, 144, 146, 155; Círculos de Aprendizaje, 10, 85–97, 99–106, 139–145; Institución Educativa Madre, 90–92, 143; scaling-up, 10, 85, 86, 88–95, 99–105, 136, 139, 143–146

Ferguson, J. 109, 115, 123
Fowler, A. 48, 132, 137, 138
Fundación Escuela Nueva Volvamos a la Gente (FEN), 10, 85–90, 103–5, 145

Garretón, M. A. 76, 78–79
Gaviria, M. C. 85, 87, 90
Ginsburg, M. 67, 80, 109
Gustafson, B. 111

Hehenberger, L. 87–88
Hulme, D. 42, 44, 46, 60–61, 85, 105, 106–7, 117, 126, 136, 139, 150

International Monetary Fund (IMF), 3, 80

Jenkins, C. 6, 24–25, 44, 86, 88, 131, 140

Klees, S. 109, 117

Lewis, D. 4, 44, 129, 131–133, 137

McEwan, P. J. 67, 139
Mexicanos Primero, 7–9, 13, 18–21, 22–40, 42–43, 45, 47–59, 136, 137, 144–45; accountability, 48–60; advocacy, 25, 26, 38, 144; Aprender Primero, 25, 35–36; mission, 18, 25, 33, 39; stakeholders, 8, 9, 12, 26, 29, 33, 34, 43, 44, 45, 47–60; strategic framework, 8, 18, 19, 25, 39, 50, 52
México: Bécalos, 19–20, 40; Coordinadora Nacional de Trabajadores de la Educación (CNTE), 22, 27, 29, 35, 57, 61; education, 8, 20, 26, 29, 32, 35, 39; education system, 20, 22, 25, 28; National Teacher's Union (SNTE), 20–22, 26–28, 36, 57; Partido Revolucionario Institucional (PRI), 20–22, 57; Secretaría de Educación Pública (SEP), 20, 22, 29, 33, 41
Minkoff, R. 6, 16, 20, 24, 130

Moulián, R. 63, 68
Mundy, K. 4, 16, 39, 42, 44, 110, 134
Murphy, L. 4, 16, 110, 134

neoliberalism, 65, 108–11, 114–15, 119–20, 123–24
non-governmental organizations (NGOs), 3, 5, 9–11, 44–45, 77, 85–86, 88, 104, 105, 108–120, 123–27, 129

Organization for Economic Co-operation and Development (OECD), 2, 27, 65, 67, 69–71, 80
Organizational capacity, 8, 18, 25, 37–38, 40, 60, 134
Organizational strategies, 11, 24, 29, 33, 36, 39, 105, 138, 148
Ortiz-Lemos, A. 112–13
Oxhorn, P. 5

parents and teachers, 8, 12, 26, 33–34, 42–43, 48, 52–53, 54–56, 59–60
Peña Nieto, E. 21, 27, 57
Program for International Student Assessment (PISA), 2, 27, 65, 70

Quality education, 9, 11, 15, 18–19, 21, 23, 25, 27, 29, 33–35, 37, 39, 41, 45, 71

Salamon, L. 5–6, 23, 47, 147
scaling-up strategies 1, 11, 13, 14, 134, 138–141
Scott, R. 133
Stromquist, N. 4, 39
student movements, 1, 5, 9, 12, 63–65, 135
Sutton, M., Arnove, R., 4, 109

United Nations Educational, Scientific and Cultural Organization (UNESCO), 1–2, 4, 69, 80, 88, 142
United Nations High Commissioner for Refugees (UNHCR), 108–9, 111–13, 115, 116–24
United States Agency for International Development (USAID), 88, 142

Vavrus, F. 86, 105
Verger, A., Novelli, M. 4, 25, 39, 42, 44, 127

World Bank, 3, 68, 80, 87, 139, 146